Defense and the Environment:
Effective Scientific Communication

NATO Science Series

A Series presenting the results of scientific meetings supported under the NATO Science Programme.

The Series is published by IOS Press, Amsterdam, and Kluwer Academic Publishers in conjunction with the NATO Scientific Affairs Division

Sub-Series

I. Life and Behavioural Sciences	IOS Press
II. Mathematics, Physics and Chemistry	Kluwer Academic Publishers
III. Computer and Systems Science	IOS Press
IV. Earth and Environmental Sciences	Kluwer Academic Publishers
V. Science and Technology Policy	IOS Press

The NATO Science Series continues the series of books published formerly as the NATO ASI Series.

The NATO Science Programme offers support for collaboration in civil science between scientists of countries of the Euro-Atlantic Partnership Council. The types of scientific meeting generally supported are "Advanced Study Institutes" and "Advanced Research Workshops", although other types of meeting are supported from time to time. The NATO Science Series collects together the results of these meetings. The meetings are co-organized bij scientists from NATO countries and scientists from NATO's Partner countries – countries of the CIS and Central and Eastern Europe.

Advanced Study Institutes are high-level tutorial courses offering in-depth study of latest advances in a field.
Advanced Research Workshops are expert meetings aimed at critical assessment of a field, and identification of directions for future action.

As a consequence of the restructuring of the NATO Science Programme in 1999, the NATO Science Series has been re-organised and there are currently five sub-series as noted above. Please consult the following web sites for information on previous volumes published in the Series, as well as details of earlier sub-series.

http://www.nato.int/science
http://www.wkap.nl
http://www.iospress.nl
http://www.wtv-books.de/nato-pco.htm

Series IV: Earth and Environmental Sciences – Vol. 39

Defense and the Environment: Effective Scientific Communication

edited by

Katarina Mahutova
U.S. Environmental Protection Agency,
Seattle, WA, U.S.A.

John J. Barich III
U.S. Environmental Protection Agency,
Seattle, WA, U.S.A.

and

Ronald A. Kreizenbeck
U.S. Environmental Protection Agency,
Seattle, WA, U.S.A.

Kluwer Academic Publishers

Dordrecht / Boston / London

Published in cooperation with NATO Scientific Affairs Division

Proceedings of the NATO Advanced Research Workshop on
Effective Scientific Communication Related to Recent Environmental Protection
Challenges and Activities in the Military Sector
Bratislava, Slovak Replic
22–26 April 2003

A C.I.P. Catalogue record for this book is available from the Library of Congress.

ISBN 1-4020-2083-X (PB)
ISBN 1-4020-2082-1 (HB)
ISBN 1-4020-2084-8 (e-book)

Published by Kluwer Academic Publishers,
P.O. Box 17, 3300 AA Dordrecht, The Netherlands.

Sold and distributed in North, Central and South America
by Kluwer Academic Publishers,
101 Philip Drive, Norwell, MA 02061, U.S.A.

In all other countries, sold and distributed
by Kluwer Academic Publishers,
P.O. Box 322, 3300 AH Dordrecht, The Netherlands.

Printed on acid-free paper

Printed in the Netherlands.

Contents

Contributing Authors

John J. Barich, III, *USA*
Doina A. Botzan, *Romania*
František Božek, *Czech Republic*
Simon M. Clark, *UK*
Mihaela Pauca – Comãnescu, *Romania*
Gaynor Dawson, *USA*
Michael Dawson, *Canada*
K. C. Donnelly, *USA*
Jiří Dvořák *Czech Republic*
Wolfgang A. Grimm, *Germany*
Dmitry M. Grodzinsky, *Ukraine*
Zoltan Jaczo, *Hungary*
Irma Jakimavičiūtė
Ilgonis Klavins, *Latvia*
Aleš Komár, *Czech Republic*
Petr Kozel, *Czech Republic*
Ronald A. Kreizenbeck, *USA*
Lubomir Kušnir, *Slovakia*
Katarina Mahutova, *USA*
Marjan Malesic, *Slovenia*
Juraj Marcin, *Slovakia*
Roger T. Moore, *UK*
Jozef Németh, *Slovakia*
Katarina Pavličkova, *Slovakia*
Steve H. Safe, *USA*
Andrejs Silins, *Latvia*
Juraj Silvan, *Slovakia*
Nathan Smith, *USA*

Preface

The NATO Advanced Research Workshop, "Effective Scientific Communication Related to Recent Challenges and Activities in the Military Sector," was the result of close collaboration between defense and environmental officials in the United States and the Slovak Republic. It was recognized that effective solutions to environmental problems are achieved through the proper application of science, including the communications and the social sciences.

Workshop was co-directed by Messrs. Ronald A. Kreizenbeck of US EPA and Lubomir Kušnír of Slovak Ministry of Defense.

Workshop topics were grouped into five major topic areas including 1) Environmental Policy Perspectives, 2) Environmental Management Systems and Public Participation, 3) Environmental Security, 4) Environmental Decision Support, and 5) Protection of Nature and Natural Systems.

Delegates from nineteen countries assembled to explore these topics at the Hotel Danube in Bratislava from April 22-26, 2003. A total of forty six delegates were in attendance, twenty from NATO countries and twenty six from Partners for Peace countries. The final program included technical sessions using a presentation and dialog format, work groups, field trips, and networking sessions. There were twenty nine technical presentations divided into seven sessions, and these form the basis for these proceedings.

Acknowledgments

The organizers of this Advanced Research Workshop acknowledge the support of two major financial sponsors, the NATO Science Programme and the U.S. Department of Defense (USDOD) under the Partners for Peace Program (PfP). Together these organizations provided the majority of assistance for both Workshop organization and delegate participation. The USDOD was assisted through the system of PfP national coordinators, with special contributions from the host PfP Coordinator in the Slovak Republic.

In addition to the Advanced Research Workshop, the NATO Science Programme provided financial support for the publication of these proceedings, and for this the Co-directors and editors are extremely grateful.

Substantial additional contributions were provided by both the Ministry of Defense of the Slovak Republic (MODSR) and the U.S. Environmental Protection Agency (USEPA.)

MODSR provided logistical support before, during, and after the workshop. They hosted the opening banquet, organized field visits to significant military and environmental sites, provided local transportation support to delegates, and full office services during the duration of the workshop.

The Slovak youth folk group, Kozlúbek, offered a program of traditional Slovak folk music, both at the opening banquet and at the official welcome by the Lord Mayor of Bratislava.

Dr. Juraj Silvan, Executive Director of Za Zdrave Životné Prostredie, managed sponsor funds, negotiated all needed commercial services, and designed the social program.

A Research Scholar with the USEPA, Ms. Katarina Mahutova, served as Workshop Manager. USEPA organized the overall program, prepared all pre-workshop materials, conducted the publicity program, and managed all activities throughout the five days of the workshop.

John J. Barich, III

Introduction

The Ministry of Defense of the Slovak Republic (MODSR) and the U.S. Environmental Protection Agency (USEPA) face many of the same challenges in fulfilling their national and international environmental management responsibilities. MODSR and USEPA collaborated to identify a set of technical issues common to military bases in both countries.

Effective communication of scientific results in an environment of rapid change emerged as the unifying and universal theme. Specific environmental issues related to operations at military facilities included 'encroachment' on training ranges, implementation of effective environmental management systems, pollution prevention within the military sector, and public participation, especially as it relates to the unique requirements of national security.

MODSR officials recognized that these issues were common to all newly admitted NATO countries, and that any scientific response done in conjunction with the USEPA should include all countries from the Central and Eastern European (CEE) region, as well as long-term NATO members from Western Europe.

An Advanced Research Workshop (ARW) sponsored by the NATO Science Programme, was identified as the preferred format to vett these issues. The U.S. Department of Defense agreed to co-sponsor the workshop in furtherance of goals of the Partners for Peace (PfP) Program. Accordingly, application was made, sponsorship secured, and an ARW convened in April or 2003.

Each of the key lecturers was invited to submit a manuscript for publication. Most chose to do so, and these manuscripts were assembled by the editors into these proceedings. They are divided into five Chapters that correspond approximately to the seven workshop sessions.

In Chapter 1, "Environmental Policy Perspectives," papers were presented that identified selected major environmental issues of interest to all militaries. *Silvan* provided a review of the major international conventions related to the environment. *Kreizenbeck* discussed environmental aspects of natural resource assets within the military sector. Delegates from the Czech Republic, Estonia, Hungary, and Poland presented papers on the environmental policy programs in their respective countries.

Chapter 2, "Environmental Management Systems and Public Participation," includes five papers related to the use of Environmental Management Systems (EMS) and Public Participation in meeting environmental policy objectives. *Clark* discusses the role of public participation in decision-making at Ministry of Defense Sites. *Kusnir* described the use of Kuchyna Air Base as a model for the Ministry of Defense of the Slovak Republic in implementing EMS requirements and other national environmental policies. *M. Dawson* provided a perspective of problems and opportunities in implementing EMSs in the military sector. This paper provided a basis for one of the three work groups that were convened as a part of this workshop. *Moore* summarized a 'corporate' EMS for the UK Ministry of Defence, and *Pavlickova* reviewed recommendations from a multi-year study on environmental impact assessement.

Chapter 3 is devoted to the contemporary topic of "Environmental Security." *Barich and Mahutova* provide an overview of EMSs, reliability management, and the risk paradigm within contemporary environmental security settings. *Silins* discusses non-ionizing electromagnetic radiation, and what this means to environmental security. *Grodzinsky* suggests that the Chernobyl catastrophe can serve as a prototype for response and effects from nuclear terrorism events. *Malesic* presented a case study of environmental security in Slovenia. And *Zadorsky* defined military conversion/restructuring as integral elements in the Ukraine's sustainable development strategy. He suggests that the Ukraine situation is not dissimilar to that faced by many other developing countries with former extensive investment in military assets.

Chapter 4, "Environmental Decision Support and Research," provided examples of research programs and tools that are available to assist environmental managers in meeting requirements of the military. *Komar* describes the development and use of an index of acceptability for military training. The index assists military managers to assess the appropriateness and potential environmental consequences of a particular training activity in a given environmental setting. *Bozek* contributed two papers on risk assessment methodologies at military facilities, and a case study on how these apply to large petroleum storage facilities. *G. Dawson* presented a risk management and remedy optimization tool for evaluating large contaminated soil sites. The methodology allows for the scale of analysis to match the needs of the issue under investigation, and for small-scale results to be passed to macro-scale evaluations, and vice versa. *Wilburn* reviewed environmental research capabilities and deployments at the Los Alamos National Laboratory in the United States, including how these capabilities might be used within the military sector. And finally, *Donnelly* presented the US Superfund Basic Research Program, a research program that stresses interdisciplinary research, training, and translation of research results into practice.

Chapter 5, "Protection of Nature and Natural Systems," includes papers related to ecology and nature conservation. *Jakamaviciute* describes the program to protect rare species and habitats as implemented by the Lithuanian Armed Forces. An innovative program to raise and protect birds of prey for re-introduction into previously damaged habitats, and to use them as a natural control for nuisance bird populations at Slovak military facilities was presented by *Marcin*. *Grimm* reviewed the experience of the US Army in protecting natural habitats in Europe. And *Pauca-Comanescu* and *Botzan* presented a case study on Romanian shrublands as an ecosystem threatened by conversion to higher economic uses, and as a valuable training setting for the Romanian military.

Three work groups all explored in facilitated discussion format the same basic question: "what are the basic communications needs as they apply to issues that are critical to the military in fulfilling its environmental responsibilities." Issues were selected based on interests of the delegates and include:

- Work Group 1: What is 'good science' and how does it apply to environmental management within the military sector?
- Work Group 2: What influences the rate of adoption and quality of Environmental Management Systems within the military sector?
- Work Group 3: What role does/should public participation play in military environmental management decisions?

In general, work group discussions added an added practical perspective to the papers published in these proceedings.

'Good science' requires careful research investment strategies, quality management, and peer review. Research investments benefit when science needs are matched to operational military requirements, and when competitive processes are used to allocate resources. Competitive proposals are reviewed from both excellence-of-science and relevancy perspectives. The former requires qualified peer reviewers while the latter requires experienced operational users of the existing science.

Quality management includes the proper use of quality management tools such as the careful establishment of data quality objectives and having recognized quality control procedures as an integral part of the execution of every science project. The most effective quality management and quality control plans are those that are documented in writing following well-known procedures. External, preferably peer, review of draft documents assures that the quality process contributes to the realization of 'good science.'

Peer review of science results is an identifying characteristic of 'good science.' The timing and extent of peer review is most effective when planned during the early stages of any project. Sufficient time and resources should be available to both obtain the peer reviews and to respond to critical observations.

The rate of adoption and quality of Environmental Management Systems (EMS) at military facilities requires support and leadership from the highest levels of management, adequate resources to both develop and implement EMSs, and acknowledgment by Commands that personnel involved in the EMS process are an integral and important part of facilities and operations management. In addition, because of the rapid development of EMS tools and techniques, it is important to provide for information and experience exchange among NATO member, PfP, Mediterranean, and other countries. This is a formative period for EMSs in military settings. Future successes will depend, in part, on the responsiveness and robustness of the initial systems, as well as the enthusiastic acceptance by facility and operations managers.

Public participation in environmental decision making in military settings must balance national security objectives with the public's right-to-know. The PfP countries and the new NATO members are all developing their public participation traditions, many guided by the principles of the Aarhus Convention. Work group members stressed the need for military public participation case studies, and information and experience sharing.

PART I:

ENVIRONMENTAL POLICY PERSPECTIVE

International Conventions in Environment: For Better Mutual Co-operation

Juraj Silvan
UNEP Basel Convention Regional Training Centre in Bratislava, Slovakia

INTRODUCTION

The goal of this paper is to briefly analyze the eight most important environmental international conventions (protocols and amendments) under auspicious of UN from the point of view of their most effective implementation. Each of the eight conventions is guided by their own secretariat. In practical life of individual countries, of course, only a small number of specialists are connected with their everyday application. Within the military sector, the responsible staff for whole environment consists sometimes only from very small sub-departments of one or two experts. Therefore it is necessary to think about possibility of bringing more work effectiveness to the implementation structures of environmental sectors and of these conventions. It is necessary to avoid the overloading of environmental experts and focus them on goal oriented works, for cooperation and networking at all levels. Good examples are The Basel Convention, CITES and Montreal Protocol for prevention, monitoring and combating of illegal traffic.

PURPOSE /AIM OF THE PAPER

The purpose of the paper is as follows:

- to analyze (briefly) eight most important environmental international conventions under auspicious of United Nations from the point of view of their effective implementation;
- to find common methodological features of these Conventions which would be able to increase their better understanding and implementation in the military sector practice;
- to draw attention to the necessity for closer cooperation and networking at all levels (within a country and internationally), among all experts and actors (including citizen participation) who contribute, institutionally and non-institutionally, for sustainability of the life on our Blue Planet.

K. Mahutova et al. (eds.), Defense and the Environment: Effective Scientific Communication, 3–14.

EIGHT UN ENVIRONMENTAL CONVENTIONS – BRIEF REVIEW AND THEIR COMMON FEATURES

I. UN ECE Convention on Long-Range Transboundary Air Pollution and its Protocols

This Convention is an essential legal framework for controlling and reducing the damage to human health and the environment caused by transboundary air pollution.

In 1972 the Stockholm UN Conference on Human Environment signaled the start of active international cooperation to combat acidification. Several studies confirmed the hypothesis that air pollutants could travel several thousands of kilometers before deposition and damage occurred.

In 1979 the Convention was signed by 34 Governments and the European Community (EC). The Convention set up an institutional framework bringing together research, policy development and implementation. In 1983, it became enforceable and today approximately 50 parties have ratified it.

Protocols:

1984 Geneva Protocol on Long-term Financing of the Cooperative Programme for Monitoring and Evaluation of the Long-range Transmission of Air Pollutants in Europe (EMEP)

1985 Helsinki Protocol on the Reduction of Sulphur Emissions or their Transboundary Fluxes by at least 30 %

1988 Sofia Protocol concerning the Control of Emissions of Nitrogen Oxides or their Transboundary Fluxes

1991 Geneva Protocol concerning the Control of Emissions of Volatile Organic Compounds or their Transboundary Fluxes

1994 Oslo Protocol on Further Reduction of Sulphur Emissions

1998 Aarhus Protocol on Persistent Organic Pollutants (POPs)

1999 Gothenburg Protocol to Abate Acidification, Eutrophication and Ground-level Ozone

II. UNEP Convention on International Trade in Endangered Species of Wild Flora and Fauna (CITES)

Because wildlife (wild animals and plants) crosses international borders, the effort to regulate them requires international co-operation to safeguard certain species from over-exploitation. CITES was conceived in the spirit of such co-operation. First, a Treaty was drafted in 1963; then, in 1973, the text of the Convention was agreed upon by 80 countries in Washington D.C.; and, in 1975, CITES became enforceable.

The aim of CITES is to ensure that international trade in specimens of wild animals and plants does not threaten their survival. The Convention combines wildlife and trade themes with a legally binding instrument for achieving conservation and sustainable use objectives.

Although CITES is legally binding on the Parties it does not take the place of national laws. Rather it provides a framework to be respected by each Party, which has to adopt its own domestic legislation to ensure that CITES is implemented at the national level.

Species listed in CITES are divided between three appendices:

- Appendix 1 includes species that are threatened with extinction. International trade is generally prohibited.
- Appendix 2 includes species not necessarily threatened with extinction, but for which trade must be controlled to avoid their becoming threatened. International trade is permitted but controlled.
- Appendix 3 includes species for which a country is asking Parties to help with its protection. International trade is also permitted but controlled.

Since 1975 not one species protected by CITES has become extinct as a result of trade. Today, it accords varying degrees of protection to more than 30,000 species of animals and plants and is supported a total of 155 Parties.

III. UN ECE Convention on Environmental Impact Assessment in Transboundary Context

The objective of this Convention is to promote environmentally sound and sustainable economic development, through the application of environmental impact assessment (EIA), especially as a preventive measure against transboundary environmental degradation. The Convention stipulates the obligation of Parties to assess the environmental impacts of certain activities at an early stage of planning. It also lays down the general obligation of States to notify and consult each other on all major projects under consideration that are likely to have significant adverse transboundary environmental impacts.

Parties to the Convention are under an obligation to take policy, legal and administrative measures to control adverse transboundary impacts arising from proposed activities. The Convention also provides conditions for research activities among the Parties, for the purpose of improving methods of EIA, and promoting sustainable economic activity. A dispute settlement process is also contained within the Convention.

IV. UNEP Stockholm Convention on Persistant Organic Pollutants

The Stockholm Convention is an international agreement aimed at protecting human health and environment from persistent organic pollutants (POPs). POPs are carbon-based compounds that remain in the environment for long periods, become widely distributed geographically, accumulate in the fatty issue of living organism, and are toxic to humans and wildlife.

Because of these properties, POPs cause damage wherever they travel. Their effects on humans and wildlife include birth defects; cancer; and dysfunctional immune, development and reproductive systems. The 12 POPs initially covered by the Convention are:

- aldrin, chlordane, DDT, dieldrin
- dioxins, endrin, furans
- heptachlor, hexachlorobenzene
- mirex
- PCBs and
- toxaphene.

In addition to the UNEP Stockholm Convention on POPs, the UN ECE Convention on Long-Range Transboundary Air Pollution has developed a protocol to deal specifically with POPs. The UN ECE POPs Protocol was adopted in 1998 in Aarhus (Denmark). It focuses on a list of 16 substances that have been singled out according to agreed risk criteria. These substances comprise ten pesticides (aldrin, chlordecone, chlordane, DDT, dieldrin, endrin, heptachlor, HCH/lindane, mirex and toxaphene), three industrial chemicals (hexachlorobenzene, hexabromobiphenyl and PCBs) and three by-products/contaminants (dioxins, furans and PAHs).

The ultimate objective of the Protocol is to eliminate any discharges, emissions and losses of POPs. At this moment the Protocol bans the production and use of some products outright, others are scheduled for elimination at a later stage, and finally the Protocol severely restricts the use of DDT, HCH (including lindane) and PCBs.

The Protocol includes provisions for dealing with the waste of products that will be banned. It also obligates Parties to reduce their emissions of dioxins, furans, PAHs and HCB below their levels in 1990 (or an alternative year between 1985 and 1995). For the incineration of municipal, hazardous and medical waste, it lays down specific limit values.

Since 2001 the Convention has been adopted by 127 governments. It will come into effect when 50 countries have ratified it; in the meantime, Party countries have agreed to voluntarily implement the provisions of the Convention on an interim basis.

Countries that implement the Convention will be required to stop the production and use of the POPs. There are provisions that allow countries to continue to use POPs for limited agreed upon purposes such as public health and the use of DDT for vector control.

V. UN ECE Convention on the Transboundary Effects of Industrial Accidents

The Convention aims at protecting human beings and environment against industrial accidents by preventing such accidents as much as possible, by reducing their frequency and severity, and by mitigating their effects. It promotes active international cooperation between the contracting Parties before, during, and after an industrial accident. It also encourages Parties to help each other in the event of such an accident, to co-operate on research and development, and to share information and technology.

The Convention was signed by 26 UN ECE member countries and the European Union and entered into force in 2000. The point of contact network now comprises 35 countries and the European Union.

The Convention has five key requirements for Parties:

- prevention (reduction of risks)
- preparedness (no matter how strict the safety standards, accidents will occur and countries must be prepared to deal with their consequences. Public awareness and participation is regarded as a key component of successful preparedness)
- response (to minimize the accidents effects, including those of a transboundary nature)
- notification (coordinated way if an industrial accident occurs; Parties must be informed as soon as possible. It is necessary to set up special notification systems, warning system, etc.)
- competent authorities and points of contact.

VI. UNEP Biodiversity Convention and CMS (Conservation of Migratory Species and Wild Animals) Agreements

In 1972 it was first recognized that many threatened wildlife species migrate long distances and cross international borders many times throughout their lives. As a result, a process was initiated to develop an international convention to protect and conserve these animals throughout their entire migratory range.

The Convention includes two appendices which list migratory species that would benefit from conservation measures taken by the Range States. Range States are those countries in which the migratory species occur. It would be prudent if the Range States worked towards improving their conservation status by means of concluding regional agreements, and declaring the protection and sustainable use of these species to be the objective of their nature conservation policy. A number of such Agreements are in effect. For example, treaties have been concluded on the conservation of seals in the Wadden Sea, bats in Europe, small cetaceans in the Baltic and North Seas and the Agreement on the Conservation of African-Eurasian Migratory Waterbirds.

This Convention was signed in 1979 in Bonn (Germany) and in 1983 came into force as an international law.

The second major Convention to attempt to address the loss of species and ecosystems was conducted in 1992: The Convention on Biological Diversity, one of the main outputs of the UN Conference on Environment and Development in Rio de Janeiro, Brazil, was the first global agreement on the conservation and sustainable use of biological diversity.

The Convention is comprehensive in its goals. It has three main goals: (1) conservation of biodiversity; (2) sustainable use of the components of biodiversity; and (3) sharing the benefits arising from the commercial and other utilization of genetic resources in a fair and equitable way.

Over 150 governments signed the document at the Rio Conference and since then more than 175 countries have ratified the agreement.

VII. UN ECE Convention on Protection and Use of Transboundary Watercourses and International Lakes & its Protocols

The Convention is intended to strengthen national measures for protection and ecologically sound management of transboundary surface water and ground-waters. It obligates Parties to prevent, control and reduce water pollution from point and non-point sources. It also includes provisions for monitoring, research and development, consultations, warning and alarm systems, mutual assistance, institutional arrangements, and the exchange and protection of information, as well as public assess to information.

VIII. UNEP Basel Convention on Transboundary Movement of Hazardous Wastes and their Disposal

The Convention is the response of the international community to the problems caused by worldwide production of wastes which are hazardous to people or the environment because they are toxic, poisonous, explosive, corrosive, flammable, eco-toxic, or infectious. The Convention grew from concerns over shipments of hazardous wastes from industrialized countries to developing countries. Principally it was devoted to setting up a control system for transboundary movements of hazardous wastes, based on prior written notification.

It provides a framework for identification, notification, control and environmentally sound management of hazardous wastes. The key objectives of the Basel Convention are that:

- hazardous wastes should be treated and disposed of as close as possible to the places where they are produced
- transboundary movements of hazardous wastes should be reduced to a minimum consistent with their environmentally sound management
- hazardous waste generation should be reduced and minimized at source, i.e. the necessity to introduce and develop clean production technologies as much as is possible.

The Basel Convention rests on two pillars: the notification and control system; and hazardous wastes management. Both pillars depend on each other for full implementation of the Conventions objective.

The Convention was adopted was adopted in 1989 and entered into force in 1992. Approximately 150 countries and the European Union are Parties to the Basel Convention.

The Basel Ban Amendment

The Basel Ban is an export ban for OECD countries, i.e. the OECD countries may not send hazardous waste to non-OECD countries for either disposal or recycling. Imports of hazardous waste from non-OECD countries will not be affected by the Amendment.

The Ban Amendment will enter into force when it has been ratified by three quarters (62 Parties) of the Parties present at the time of adoption of the Amendment (1995).

So far approx. 25 Parties and the European Union have ratified. The European Union has enacted legislation transposing the Ban Amendment which is binding for all 15 EU member states.

COMMON FEATURES OF THE EIGHT UN CONVENTIONS

In spite of the fact that the eight UN conventions are not all of the same type, it is possible to find the following common characteristics:

- Each Convention has important implications as well as consequences for national, international and global environmental management and protection.
- All Conventions share precautionary principles and principles of prevention. Each Convention is therefore an important instrument of conflict prevention.
- Many of the mentioned Conventions focus on transboundary issues, making implementation and in particular enforcement and border control, a cross-cutting issue.
- The issue of civil liability is central for all transboundary impacts, but it has not been elaborated in all Conventions. However, there is increasing pressure to address it in the future.
- Mechanisms for settling disputes are also addressed by the majority of the Conventions. These provisions are important for promoting security and regional co-operation.
- An important common element for all Conventions is very active public participation. This is a key element of success for any Convention both at the national and international/transboundary context. This principle has been fully elaborated and anchored in one of the recent environmental instruments, the Aarhus Convention.
- All Conventions share the need to establish inventories. Taking stock of sources of pollution **and** of hazardous installations that could cause accidents should be done in a coordinated way in order to create synergies for reducing impacts.
- All Conventions include reporting requirements where future joint, coordinated capacity building activities have to be undertaken.

LOOKING AFTER COMMON METHODOLOGICAL FEATURES: INDIVIDUAL STEPS FOR BETTER UNDERSTANDING AND CO-OPERATION

Three categories of steps are necessary to implement for useful, effective and co-operative managing of the UN Conventions:

A. FIRST CATEGORY OF STEPS:

- Consider adherence to the Conventions, their Protocols and Amendments
- Prepare national legislation for adherence
- Designate National Focal Point(s) to act as a clearing house (for collection and exchange information)
- Promote deeper understanding of philosophy and practical implementation of the Conventions′ subjects, their benefits and successes
- Move quickly for their ratification

B. SECOND CATEGORY OF STEPS:

- Develop national development documents (e.g. national implementation plans), which involve:
  - assessing the sources;
  - assessing the impact of individual sources;
  - identifying the infrastructure available to manage the issue;
  - determining additional required capabilities.
- Increase the quality of inventory, as well as reporting activities
- Increase the quality of data gathering in accordance with Conventions requirements (for air quality, enhance of number and quality of collecting data; e.g. the most frequent are reports only on sulphur and NO^x emissions)
- Contribute to the of reduction and ultimate elimination of the Conventions′ subjects (e.g. of illegal traffic of listed plants and animals – CITES, hazardous waste – Basel Convention)
- Participate at the training activities for experts (workshops, seminars, conferences)
- Participate in the development of software tools for reporting, evaluation of data, their expert assessment

- Join the effect related monitoring programmes
- Make a continuous effort for actualization of the Conventions subjects′ future priorities
- Provide Conventions with an improved and secure financial mechanism (e.g. to use financial advantages to signatory countries that are eligible for financial assistance from GEF) and administrative basis, as well as conditions for access to technical assistance

C. HIGHER GENERATION OF STEPS, "META-STEPS":

- Strengthen scientific basis of the decisions-making process
- Increase international co-operation among contracting Parties (in the field of research, development, sharing information, technology, new projects, common programmes, e.g. compatibility of plans for risk management, prevention of industrial accidents, including an early warning system)
- Conclude strategic alliances with national and international stakeholders (funding!)
- Contribute to progress towards global membership of the Conventions subjects′ and their philosophy and proper implementation
- Contribute to extension of clearing house networking
- Contribute for explanation of synergy among individual Conventions, Protocols and programmes
- Introduce clean production (clean technologies) in the whole manufacturing sphere.

CONCLUSIONS

A. MULTIDIMENSIONAL CHARACTER OF THE CONVENTIONS

- Environmental
- Technical
- Economical
- Social
- Cultural-ethical;

B. COMMON GENERAL GOALS OF THE CONVENTIONS

- Better environment
- Sustainability for existing ecosystems
- Sustainable development for future generations

It is necessary to reach closer cooperation among all international Conventions. This is the crucial issue for further more effective work on all levels.

Implementation of all above issues requires further examination of appropriate coordinated mechanism on all levels, namely at institutional, organizational, theoretically-methodological and application level. The human approach should have the leading position by all activities.

The environmental agenda, as we can see from the Conventions′ content, is an extraordinary extensive issue, diversified and demanding for knowledge, technical, as well social capability and information level. Environmental management issues are necessary in military sector using the highest priorities and appropriate consecutive steps.

C. ENVIRONMENTAL TASKS IN MILITARY SECTOR

Within the framework of military sector, it would be useful to consider:

- Strengthening the position of environment as the priority on all commanding levels: the situation regarding professional staff has, generally speaking, a lot of reserves: the environmental divisions at top military levels are need to be enhanced both in the number as well as the specialization of experts. There needs to be highly environmentally- educated staff. After successful implementation of these requirements, it would be possible to manage in a more appropriate way the whole range of the Conventions for all mankind.
- Specialization among experts is inevitable and leads to more effective co-operation with environmental sector at civil sphere on all levels.

REFERENCES

1) Workshop on UNEP and UNECE Environmental Conventions in the Federal Republic of Yugoslavia, Belgrade 14-16 November 2001 (Information kit and Summary report)

2) UNEP Regional Training Seminar for Port Enforcement Officers, Constanta, Romania 5-9 November 2001

3) RIO+10 Report. The National Report of the Slovak Republic on Sustainable Development. Editor: Ministry of E the Environment of the Slovak Republic. Compositors: Klinda, J., Bebej, J., Toma, P.; 2002

4) Silvan, J.: On Multidimensional Aspects of the Basel Convention. In: Newsletter No. 6 of the Regional Training Centre for Implementation of the Basel Convention Bratislava, Slovak Republic; September 1998
5) Lacuska, M., Silvan, J.: Slovak Regional Training Centre After Two Years of Operation. In: Newsletter No. 8 of the RTC for Implementation of the Basel Convention Bratislava, Slovak Republic; May 1999
6) Bakken, P.: Past Achievements and Future Challenges for the Basel Convention. In: Newsletter No. 9 of the RTC for Implementation of the Basel Convention Bratislava, Slovak Republic; November 1999
7) Silvan, J.: New Publication: CLEANER PRODUCTION (12/1999). In: Newsletter No. 10 of the RTC for Implementation of the Basel Convention Bratislava, Slovak Republic; March 2000

Environmental Aspects of Managing of Natural Resource Assets within the Military Sector

Ronald A. Kreizenbeck
U.S. Environmental Protection Agency, Seattle, WA. USA

INTRODUCTION

The natural and environmental resource management issues faced by military managers, the similarities and differences with other sectors, and the tools available to each are discussed in the context of competition and stewardship. Military readiness depends on the availability of an adequate natural resource base for a variety of logistic and training purposes. As populations and economies grow, competition for natural resources, and the ability to increase or even maintain the military resource base is a major challenge. This challenge is shared by all owners/managers of natural resource assets, be they military, civilian, commercial, or public. Encroachment is a term commonly used within the U.S. military sector (but not within other sectors) to describe the totality of environmental requirements faced by military resource managers. Encroachment derives from consistent application of national environmental laws and regulations, from the need to use the most current products or science and technology, from competition for limited resources, and from limits made obvious through integrated risk assessment.

DISCUSSION

Historical Perspective and Common Challenges

The modern era of environmental management in the United States began with passage of the National Environmental Policy Act of 1969 (Public Law 91-190, or NEPA), and with the creation of the U.S. Environmental Protection Agency in 1970 (USEPA). The creation of USPEPA brought together in one independent agency 15 bureaus, offices, and administrations that heretofore had been scattered throughout the U.S. Government. For the first time, governmental officials responsible for water pollution, air quality, solid waste, drinking water, pesticides and so on, were challenged with the task of managing to the goal of 'one-environment.' Goals associated with the creation of USEPA included establishing the environment as a national

K. Mahutova et al. (eds.), Defense and the Environment: Effective Scientific Communication, 15–19.

priority, avoiding solving one environmental problem (for example, industrial water pollution) by creating other environmental problems (for this example, improper management of toxic wastewater treatment sludges), achieving management efficiencies, and facilitating communications with stakeholders with an interest in the environment.

Concurrent with the creation of the USEPA, the nation's basic environmental laws including the Clean Air Act (1970), the Federal Water Pollution Control Act (1972), and the Resource Conservation and Recovery Act (1976) were updated and modernized. Each built on experience gained with predecessor laws dating back, in some cases to the 19th century, and to proven innovations in other, more recently enacted laws in other environmental media. Several principles emerged, and these are now common to every environmental statute:

- the 'polluter pays' principle, the business, organization, or individual responsible for creating a problem is responsible for correcting it,
- the 'level playing field' principle where everyone within the public and private sectors are required to comply with the same set of requirements, thereby preventing situations an organization that chooses to ignore requirements achieves by this act an unfair competitive advantage,
- the 'states rights' principle where implementation of environmental laws is encouraged at the local and State level, with the federal government (USEPA) available to support whenever the other levels of government are not discharging their responsibilities,
- the 'public involvement' principle that requires maximum opportunity for public input into environmental regulations and permits,
- the 'enforcement' principle, a set of uniform regulatory requirements that apply to all equally, and for which adequate investigative and enforcement resources are available to ensure compliance,
- the 'good science' principle where resources are provided for a national research program, and administrators are required to include the most current scientific understandings in all new requirements and decisions, and
- the 'training' principle where resources are made available to train the next generation of environmental managers.

By the 1990's, well-established environmental ministries were operational in almost all NATO countries and candidate countries. Many of these followed to some extent the USEPA model, and several of the principles (polluter pays, public involvement, enforcement, and good science, for example) are common to national legislation and international conventions.

The first common environmental challenge faced by all NATO members is how to best respond to the environment laws, regulations, and norms as a priority while still achieving the basic mission of security/readiness.

During the years of NATO's existence, there has been an ever accelerating increase in the number of commercial and industrial chemicals, pesticides, and technologies developed by both the military and domestic commercial sectors. Achieving mission goals and remaining competitive requires that organizations avail themselves of these products of modern science. But they also present new challenges for the environment.

The second common environmental challenge faced by all NATO members is how to select and deploy the products of modern science without unnecessarily degrading the environment.

Competition for limited natural resources affects both the military and the civilian sectors. The basic factors are well-known and include:

- increasing population,
- increasing demand (in general) for the benefits, the goods and services of modern society, and a concurrent demand for a healthy and pleasant environment,
- limited quantity of resources such as land, water, and air,
- depletion of fixed resources such as minerals or petroleum,
- utilization of renewable resources at a rate greater than the sustainable rate

Because military sector activities require a large 'foot print' compared to many other civilian sector activities, because military activities (especially training) require a large separation from civilian activities, and because the military sector tends to have large natural resource assets (especially land) under management, the competition for resources is an especially important challenge for military environmental managers.

The third common environmental challenge faced by all NATO members is how to properly manage the natural resource asset base in the face of increased competition for limited natural resources.

Finally, the science of risk assessment has developed rapidly in recent years, and it is becoming more evident how various military and civilian activities effect the environment, in ways that may not be anticipated by the environmental laws, regulations, and norms. In its broadest application, integrated risk assessments (IRAs), includes the examination of not only the relationship between environmental stress and its consequences (such as human health and ecological disturbance), but also how the consequences of environmental stress may influence other concerns such as social, economic, welfare, and security. Environmental managers must strive to achieve goals 'beyond compliance,' as guided by integrated risk assessments, or risk creating tomorrow's environmental problems today.

The fourth common environmental challenge faced by all NATO members is to act on the insights provided by integrated risk assessments.

Unique Potential of Military Lands

The military sector requires large tracts of land for operational and training purposes. Certain of these assets need to be co-located with areas of commerce and population. Others must be isolated so as to avoid negative impacts on adjoining land owners. These natural resource assets are often the earliest withdrawal of lands for special purposes. Withdrawals are done to meet current and future, distant future needs. An unplanned consequence of early, large land tract withdrawals is that military holdings can be important sanctuaries for unique ecosystems and threatened or endangered species. The spotted owl provides a good example of the potential and the challenges of endangered species on military lands.

In the United States, the Endangered Species Act (ESA) is the primary national law that provides protection to threatened and endangered species. The nearest international equivalent are the Convention on Biological Diversity (CBD) and the Convention on International Trade in Endangered Species (CITES), and the Bonn Convention on Migratory Species (CMS.) ESA protected species are listed in the CATALOGUS. Lists of Conventions protected species are found in a number of references, including Appendices I, II, and III of CITES, and the World Conservation Union (IUCN) endangered and threatened species database, commonly referred to as the 'Red List.'

The Pacific Northwest region of the U.S. has several important species listed as endangered, including the spotted owl, an owl whose preferred habitat for nesting includes 'old growth' forest ecosystems dominated by the Douglas fir. Douglas fir is one of the most important commercial tree species for the forest products industry, with old growth trees being the most highly valued. The industry is extremely important to the northwest for historical, geographic breadth of activity, number of jobs, and overall contribution to the regional economy. The primary threat to the spotted owl is from loss of preferred habitat and by encroachment of incompatible activities. Owners and managers of remaining spotted owl habitat must now incorporate how potential changes, either positive or negative, will affect these critical habitats.

Of the 10 million hectares (25 million acres) of U.S. territory managed by the U.S. Defense Department, 35,000 hectares (86,000 acres) are the Fort Lewis Military Reservation in Washington State. Nearly 65%, or 22,000 hectares (55,000 acres) is considered critical habitat for the spotted owl. Fort Lewis, therefore, is an important public asset for the protection of the spotted owl. Future uses of this acreage for more active military purposes involves a tradeoff between habitat preservation and military operational objectives.

The Fort Lewis spotted owl management challenge is repeated throughout the U.S. defense holdings. There are over 300 threatened or endangered species resident within U.S. military lands.

An equivalent situation exists within the Europe. Environmental policy of the European Union together with an approximation of environmental policy of accession countries to the EU facilitate the creation of an European ecological network of areas that are significant from the natural protection and conservation view. This network of ecosystems, protected at Community level, is called NATURA 2000.

Each Member State and Accession Country identifies the Special Areas of Conservation sites which are important for the protection of the species and habitats covered by the Directive on the conservation of natural habitats and of wild fauna and flora, and Special Protection Areas according to the Directive on the conservation of wild birds. Land owners of these must ensure their long- term preservation by integrating the socio-economic activities into a sustainable development strategy. This provides an opportunity to demonstrate how environmental concerns can be integrated into other sectors, either private or commercial.

Military real estate is not exempt. Assuming that military installations are home for important natural habitats, pristine, vital ecosystems and numerous threatened and endangered species, they may contribute to the success of NATURA 2000.

CONCLUSION

Environmental aspects of managing natural resource assets within the military sector differs by degree, but not by principle, managing natural resource assets within the civilian sector. Within the time of NATO's existence, comprehensive environmental laws, regulations and norms, and international conventions have been widely adopted. The same principles are found in most legal and administrative settings: polluter pays, public involvement and access to information, enforcement and compliance, regulations and norms apply to all (the level playing field), and a reliance on good science to guide action. Military natural resource assets, because of their size, longevity, and single owner/steward, contain many environmental and ecological assets of national importance. This intensifies the competition between preservation and alternative use. Environmental managers in NATO member and candidate countries face several common challenges including: effective response to laws, regulations and norms, deployment of new technology in a manner consistent with environmental stewardship, fulfillment of military operational objectives (and preserving the ability to do so into the future) in light of increasing competition for natural resources, and 'beyond compliance' management to take advantage of integrated risk assessments in a way that avoids future problems.

Because of the similar nature of environmental needs between countries, and because of the similarities in the legal, administrative, and scientific responses, experience sharing and information transfer provides an efficient means for environmental managers to meet their mission requirements.

Environmental Policy of the Czech Ministry of Defence

Petr Kozel
Environmental Department, Ministry of Defense of the Czech Republic, Prague

INTRODUCTION

Over the last decade, the achievements of the Czech Ministry of Defence (MoD) in the area of environmental protection (EP) and development has been nationally significant. In general there are some problems that have not been tackled, or were not dealt with properly for many years, especially due to a lack of funds. Indeed, less apparent technical and technological EP mistakes and shortfalls of decision-making may soon become because of increasing community pressures on the environment and its elements.

In 1999, the first Environmental Policy of the MoD created conditions favourable for correcting the existing shortfalls, implementing new precautionary measures, and for preserving and perhaps enhancing the Ministry's lead over the civilian sphere. The achievements have been in a number of environmental areas, often under arduous personnel and financial conditions. The Environmental Policy (EP Policy) was developed in accordance with National Environmental Policy. It respects commitments and obligations pursuant to the Czech Republic's membership in NATO and provided for introduction of Environmental Management Systems (EMS) in accordance with ES ISO series 14 000.

On the ground of the adoption new State Environmental Policy of 2001, the new Environmental Policy of the MoD proposal was elaborated as a contribution to Reform of the Armed Forces of the Czech Republic.

The fundamental principle of both the State Environmental Policy of 2001 and the Czech MoD Environmental policy proposal of 2002 is the principle of sustainable development.

MINISTRY OF DEFENCE ENVIRONMENTAL POLICY OBJECTIVES

The goal of the department policy is to determine the department's priorities, main tasks in particular EP areas, and instruments to implement them. Priorities and tasks are formulated in accordance with State

K. Mahutova et al. (eds.), Defense and the Environment: Effective Scientific Communication, 21–31.

Environmental Policy so that they cover the entirety of the department's activities in harmony with the defence development concept of the country and armed forces development, but also with the requirements for compatibility and interoperability with NATO or European Union.

The approach to the environment, given by new department policy, does not consist in only passive protection in terms of resolving neglected problems, (waste disposal and damage settlement etc.) but especially in a proactive creation of the environment and damage prevention during decision-making.

For this reason it is now imperative to introduce EMS to the defence department, including a commitment to improve the environment on a sustainable basis.

MINISTRY OF DEFENCE ENVIRONMENTAL POLICY PRINCIPLES

The Fundamental Principle of the Czech Republic State Environmental Policy of 2001 is the Sustainable Development Principle, which stems from harmonisation of economic, environmental and social facets of all development objectives.

Derived principles of MoD EP policy are deduced from the fundamental principle in accordance with national policy and include community and department needs and experience. They apply chiefly to development of Internal Norm-setting Acts and choice of particular tools of department policy.

- Integration Principle

Determines that EP requirements must be an integral part of all departmental activities. The Principle is basically identical to the EMS fundamental idea.

- Prevention Principle

Measures which prevent possible endangerment or damage of environment are more efficient and cheaper than subsequent aftermath removal.

- Precaution Principle

Applies to activities and substances usage for which, we are not able to anticipate their environmental impacts based on the existing knowledge, mainly form the viewpoint of cumulative and synergy effects. The scope of activities and substances usage limitation or elimination is commensurate to the extent of ignorance or unpredictability.

- In-source Hazard Diminishing Principle

Is based on the fact that the most efficient and least expensive way to minimise negative effects on the environment is via through measures taken just at the source.

- Principle of Responsibility for the State of Environment

Is based on personal responsibility of all employees and members of the department for the condition of the environment.Commander's responsibility and individual's personal responsibility is determined by both binding legal regulations and appropriate Internal Norm-setting Acts. Possible originators of an environmental damage are fined and / or repremended.

- Subsidiarity Principle

The decision-making authority and competence in the EP area should be drawn nearest to a problem, i.e. to the lowest expert-competent decision-making level possible.

- Principle of increasing ecological awareness

Is based on a comprehensive ecological education and an every-day edification of all members and employees of the Department. The Education also includes inspection with consequent measures (allowance or penalty).

- Principle of the best technique available

Is a decision -making criteria for authorising individual activities within the Department. At the same time, the best technique available is not determined by the particular activity, but by environmental parameters.

- Substitution Principle

Environmentally hazardous activities and substances are to be replaced by less hazardous or innocuous ones. This is pre-conditioned by technological applicability and economic availability.

MAIN TASKS IN THE INDIVIDUAL AREAS OF MINISTRY OF DEFENCE ENVIRONMENTAL POLICY

The new philosophy of MoD policy, based on the principles of sustainable development, prevention and transition from resolving archaic problems to EMS, defines tasks in the following main EP areas:

Concepts, Organisation and Institutional Area

- **EMS introduction into the Defence Department including all responsibilities and rights implied by EN ISO series 14000 and establishment of a group for introduction and later control of EMS within the Department.** To integrate an authority with a competence appropriate to its significance into the controlling structures of the Department. To make use of the CCMS/NATO "Environmental Management Systems in the Military Sector" pilot study, as well as of two EMS NATO/CCMS Conferences of 1999 and 2001.
- **Establishment of MoD Council for Sustainable Development in accordance with National Environmental Policy.**
- **Creation of conditions favourable for compatibility and interoperability with other NATO countries in the environmental area.** To utilise especially "Statement of Environmental Policy in NATO Armed Forces", "Environmental Education Principles" and STANAG 7174 EP.
- **Establishment of environmental emergency system.** To co-ordinate its development with the plans for the integrated rescue system and make use of the information in CCMS/NATO "Environmental Security and International Context" pilot study.
- **Development of Environmental Information System (EIS) and making the work more effective by garrison Basic Ecological Cards.**
- **Evaluation of Department's planned activities and development programs from environmental impact viewpoint prior to approving or conducting them.**
- **Determination of competence and responsibility for EP area within the Department.** To determine guarantees and stable procedures for the cases of unclear competence.
- **Completing of departmental environmental network as to the personnel, and its stable integration into MoD and General Staff structures, in accordance with NATO standards.** To develop (to complete) a network of military environmentalists, commensurate both by achievements and appropriate numbers and position in the organisation structure to other NATO-member countries. To accommodate the state administration in the EP area by regional environmental authorities to the new territorial arrangement of the state.
- **Development of a scientific research installation within the framework of environmental authorities within the department.** To analyse the technologies and materials used from the viewpoint of their compliance with EMS conditions. To share the most up-to-date information form the military EP branches in the military sector in a form of edification and education (see below) and to introduce it into

everyday life. To submit technological and conceptual proposals and comments to specialist authorities. To utilise NATO means for projects implementation as possible.

- **Renewal of Environmental Committees' mission and significance at garrisons.** To enlarge their jurisdiction with the function of Councils for sustainable development in accordance with State Environmental Policy.

Legislation Area

- **Development of a system of new environmental Internal Norm-setting Acts (INAs) reviewing of the existing ones on the basis of determined schedule or current needs.** During INAs reviewing and development to abide by its compatibility with regulations of CR, EU, NATO and EMS and sustainable development principles.
- **Thorough application of the duty to negotiate newly adopted INAs with environmental authorities of the department.**

Financial, Economy and Logistics area

- **Development of EP funding proposal in accordance with planned CR EU membership.** To co-ordinate and make more efficient using of funds for EP.
- **To make more transparent and to simplify the integration of EP-related budget items into the MoD budget composition.**
- **As a priority, to focus investment activities on liquidation of neglected problems** (see above) **and do away with prevailing environmental hazards.**
- **Elaboration of a system of material responsibility to protect the environment.** To stimulate economically feasible environmental approaches and visibly hold liable for lack of environmental discipline.
- **Analysis of suitability of technologies and materials used or planned for use, from the viewpoints set by EMS conditions.**
- **Registering and codification of buildings and installations used by the Department as for their environmental impact and effect on human health.** This concerns especially some military bases, education and training bases, equipage parks with all appliances (garages, shelters, parking lots, warehouses, POL stations, charging stations, cleaning installations), air fields, communication network, signal system, military materiel loading/unloading/transit stations, repair shops, warehouses, lodging installations and buildings to include boiler rooms, boarding facilities, waste disposal sites, military medical installations and veterinary installations.

1. To focus on environmental aspects of their operation and to gradually adapt technical or technological equipment to conditions stemming from this policy, valid legislation , INAs, commitments arisen due to EMS introduction and NATO and EU membership. To resolve postponed, neglected, and non-professionally or inconsistently tackled problems (see above). The EP in the Department cannot be narrowed down to these issues only.

2. Focus on the development of healthy working, lodging and boarding environment for members and workers of the Department.

Training Activities Area

- **Environmental optimisation of training activities to include enhancing of simulated training and virtual reality portion.**
- **Paying special attention to the territories of military training areas from the viewpoint of military-environmental solution of possible interests clash.** To finalise the countryside environmental registering and complete or innovate territorial plans of military forests so, that conditions are created for meeting both the growing requirements for training volume and EP requirements.
- **Ensuring environmental safety during exercises of foreign troops.** To abide by EP-related rules to determine the responsibility for environmental damages in negotiating agreements on other countries' troops' participation in military exercises held on the CR territory.
- **Ensuring environmental safety during external exercises and operations**

Area of Protection of Individual Elements of the Environment and the Countryside

- **Figuring out direct and indirect measures with regard to national environmental policy and the above-mentioned Defence Department Environmental Policy**. To elaborate and specify expert activities from the general, time, material and monetary viewpoint during systematic development, protection and improvement of the following elements of the environment and nature and countryside:

  - **soil and rock environment**
  - **water**
  - **air**
  - **living organisms (threatened and protected plants and animal species)**

- **nature and countryside as a whole (protected areas, landscape character, forests, agricultural areas etc.)**
- **historical cultural sights**

Area of Waste and Hazardous & Detrimental Substances Treatment

- **Completion of an integrated departmental waste treatment and registering them in accordance with the existing legislation.**
- **Harmonisation and specification of hazardous materials handling (usage, transportation, storage) from EP viewpoint.**
- **Precautionary approaches in the area of waste management**. To minimise waste production through selecting suitable materials and substances and by suitable procedures and approaches.

Area of Protection against Detrimental Factors of the Environment

- **Monitoring and protection against hazardous physical, chemical and biological factors.** To augment activities related to the radon programme with protection against adverse effects of other factors of physical pollution. To innovate the appropriate Order of the Minister of Defence in this sense. Also to tackle specific local problems in cooperation with the scientific and research base of the Department to engage in little pursued, not monitored, unknown or undocumented effects.
- **Completion of prevailing hazards liquidation.**

Inter-departmental Cooperation Area

- **Cooperation and co-ordination of some activities with the Ministry of Environment and its authorities and institutes (Czech Environmental Inspection, Hydro-meteorological Institute, Institute for environmental policy etc.)**
- **Cooperation and co-ordination of some activities with state government and self-government authorities.**
- **Cooperation with other departments (MI, MIT, MFA etc.)**

International Cooperation Area

- **Cooperation with NATO Committee on the Challenges of Modern Society when developing pilot studies on environmental issues. An active participation of national**

experts of the Department in CCMS/NATO pilot studies and co-ordination of these activities within the Department.

- **An active participation in environmental events organised by EPWG and ETWG NATO, EAPC, PfP, Mil-to-Mil and foreign civilian organisations.** To participate proactively with published achievements; utilise the lessons learnt in every day practice. To make use of scientific and research potential of environmental authorities of the Department (see above).
- **Cooperation with schools and training centres abroad in ecological education by students' participation in courses and training and via studies of international students at domestic schools.** To prefer cooperation with NATO School SHAPE and military schools of NATO-member countries.
- **EP branch experience sharing with foreign defence departments experts (bilateral or multilateral).**
- **Taking part in environmental safety concepts development in international operations.**

Promotion, Information Measures and Publication Areas

- **Extending the military environmental publishing scientific, expert and popular activities within the department and beyond.**
- **Expanding internal and external awareness of departmental environmental activities.** To utilise especially commander's meetings for internal information distribution. To promote branch achievements and opportunities of education in areas covering environmental protection at military schools by the general public.
- **Establishment of "Environmental Texts Library" for the needs of ecological education and everyday requirements of Department members.** To consider the possibility to issue Environmental Bulletin.
- **Establishment of a Departmental Award (in case agreed, an Award of the Minister of Defence) for extraordinary achievements and benefit in EP.** In regular intervals (on the occasion of the Earth Day), to award individuals staffs or organisations (workers, associates, donors and suppliers) morally or otherwise, according to determined criteria.
- **Releasing the basic data on Environmental Information System (see above) in a form of departmental ecological year book.**

Education Area

- **Development of new departmental concept of environmental edification and education.** To pay higher attention to ecological education of senior officials of the Department. When implementing the mentioned activities, to utilise military colleges, scientific and research potential of the environmental authorities of the Department and the Environmental Texts Library (see above).

Inspection Area

- **Augmenting the inspection and cognition area as of the EP prevention and education method area.**

MINISTRY OF DEFENCE ENVIRONMENTAL POLICY MEANS

Treatment of the Environment elements must be acceptable not only from the ecological but also economical, social and political viewpoints within both the Czech Defence Department framework and international context.

The Defence Department Environmental Policy can be implemented while using the following set of means only:

Political vehicles consist in elaboration upon the departure political materials into partial political documents. The main internal policy tool is preparation, introduction and operation of EMS. The International policy-related ones include especially effective international treaties and agreements in the EP area, the National EU membership preparation programme and NATO environmental materials.

Organisational and institutional tools consist in a thorough implementation of the effective EP Order of MoD and its amendment in the sense of establishing Councils for sustainable development at MoD and lower control echelons, establishment of Group for EMS, development of regional information system on the environment and environmental emergency system. Consulting all activities and documents with environmental authorities must become once again an integral part of the department activities organisation. Institutionalisation of state administration in EP by MoD regional environmental authorities will be conducted in accordance with the new state territorial arrangement. Apart from the aforesaid authorities, the conceptual tools will be developed by the scientific and industrial base and the existing environmental committees.

Personnel and competence tools. The main personnel means include strengthening of environmental authorities at MoD and the GS, Formations and units, establishment of new authorities (see above) and creation of posts for graduates of environmental branches at military colleges. The competence tools consist particularly in new determination of jurisdiction, responsibilities and competence of departmental environmental authorities and extending the jurisdiction of the existing environmental committees with the function of the Councils for sustainable development.

Legislation tools are represented by Internal Norm-setting Acts and NATO standards on EP issues.

Financial and Investments tools. The EP area financial policy tools include both internal means from the MoD state budget chapter and, to a lesser extent, subsidies from Ministry of Environment's programmes, or subsidies of other authorities and organisations. Department's EP investment policy will consist in both doing away with prevailing hazards and "archaic" problems, and chiefly in new investments in support of introduction and operation of EMS.

Material and Technical tools are represented by all technical, technological and material means for implementation of EP expert activities. Their quality and suitability from environmental viewpoint will be monitored according to EMS and sustainable development principles. At the same time, the primary task is to perfect the technical and technological base from the EP viewpoint.

Information tools serve to inform both internally and externally and will be represented by the regional environmental information system and registering and codification of buildings and installations from the viewpoint of their effect on the environment while using Unit Basic Ecological Cards. The data not subject to classification can be presented through the Environmental Bulletin, Ecological Year Book and individual publications of departmental experts.

Education tools consist especially in education of workers and members of department on EP issues pursuant to the appropriate INA. They also cover an every-day edification and increasing the environmental awareness. The most effective training tools are simulated training and virtual reality.

Scientific and Research tool. The most efficient analytic and synthetic tools are the development of military ecology and practical application of research and studies achievements. Available tool for elaborating upon Defence Department Environmental Policy is also Agenda 21, or development of local "Agenda 21s".

Inspection tools consist in conducting external and internal inspections of all levels and state administration execution in EP by MoD regional environmental authorities.

Territorial Planning is one of the important EP tools especially in the military training areas.

MINISTRY OF DEFENCE ENVIRONMENTAL POLICY IMPLEMENTATION

The main tasks are to be implemented through the above-mentioned tools in various time-frames. The measures can be divided into short-term, mid-term, long-term, progressive and immediate ones according to urgency and availability of individual tools.

DEFENCE DEPARTMENT ENVIRONMENTAL POLICY VALIDITY

The document is developed for the period of time until 2006. It can be specified or re-drafted pursuant possible adoption of a new national environmental policy, or as a response to significant political and expert domestic and international initiatives.

REFERENCES

1. Ministry of the Environment of the Czech Republic: State Environmental Policy. Prague, 2001.
2. Kozel, Petr: Proposal of the Czech MOD Environmental Policy. USECOM and PfP Environmental Conference, 7-11 May, 1999, Krakov, Poland

Environmental Policy in the Estonian Defence Forces

Andres Rekker
Estonian Defence Forces Joint Headquarters, Tallinn, Estonia

INTERESTING FACTS ABOUT ESTONIA

Estonia's area of 45, 227 km2 is similar to that of the Netherlands, but the population (1.4 million) is ten times smaller. Daylight on the longest day lasts over 19 hours, while the shortest winter day lasts only six hours.

There are more than 1,500 islands, 1,000 lakes (5 per cent of Estonian territory), 7,000 rivers and streams in Estonia. Bogs and wooded swamplands of different types cover over one fifth of the country.

Estonia is a seafront country - the length of coastline (3,800 km) is about 6 times longer than the mainland borderline.

Estonia is very rich in timber; various kinds of forests cover almost half of Estonia's territory. Elks, wild boar, bear and lynx are amongst Estonia's common large mammals. About 10 per cent of Estonian territory is a nature reserve.

The average temperature of the coldest month, February, is -3.5° to -7 °C, on the other hand the temperature of the warmest month warmest, July, is 16 to 20°C.

OVERVIEW OF THE ESTONIAN DEFENCE FORCES

The Estonian Defence Forces consist a regular component (divided into three services: Army, Air Force and Navy), Military Educational Institutions and a voluntary military organisation called the Defence League.

In 2003, the average number of members in the defence forces is about 5,200, divided between components as follows:

Army	4,500
Air Force	200
Navy	400
Defence League (National Guard)	9,500

K. Mahutova et al. (eds.), Defense and the Environment: Effective Scientific Communication, 33–41.

PRESENT SITUATION AT THE MINISTRY OF DEFENCE AND DEFENCE FORCES

Estonian Defence Forces are the integral part of the society and they are required to follow the current legislation. The Peacetime National Defence Act regulates the activities of the Ministry of Defence and Defence Forces in peacetime and therefore the Ministry of Defence and Defence Forces are obliged to follow legislation which concerns the matters of protection of environment.

Since 1999, the Defence Forces regulation "Inner rolls" are unit commanders with responsibility for environmental security and protection.

Order of the Supreme Commander No. 522 (28, June 2001) and Order of the Ministry of Defence (20, June 2001) ratify the Environmental Policy Act for Defence Forces. These are based on the Act related to the Protection of Nature in Estonia and Defence laws, and lower acts, and draft STANAG 7141 EP "Joint NATO Doctrine for Environmental protections during NATO-led Operations and Training."

In the Republic of Estonia, since 1988, state laws and lower legal acts regulate environmental protection. The environmental (civilian and military) regulations concerning the Defence Forces regulate the training activities and establishing of dislocation sites, as well as their use. In addition to the above-mentioned, the Republic of Estonia has joined with several international nature protection conventions when developing the Defence Forces. In case of an environmental damage caused by the activities of the Defence Forces, they must take responsibility if it has taken place as a result of an intentional activity or negligence.

In planning their activities, the Defence Forces of the Republic of Estonia have to proceed from the possibility of causing minimum damage to the environment. They must take care of nature and preserve the diversity of nature; and at the same time, take into consideration the obligations and needs of the national defence.

In order to carry out the function mentioned above, in the autumn of 1994, the Ministry of Defence (MoD) established the post of the Environmental Expert Under State Secretary. Its obligation is to coordinate and carry out inspections and create conditions for the organizations authority of the Ministry of Defence for fulfilling the current environmental legislation and using the natural resources economically.

Since 1997, after the re-structuring of the MoD, this function is part of the Real Estate and Construction Department. This department plans and arranges construction designing, the subsequent public procurements and construction contracting in the area of government of the Ministry and guarantees the conformity with the environmental requirements. In the Defence Forces leading Environmental work for all three Services is Army Staff under Engineer and NBC-Defence

Department. This is the section dealing with the environmental problems of the Defence Forces' real estate, mostly in the training areas, in the logistics department of the General Staff of the Estonian Defence Forces

HISTORICAL BACKGROUND

Estonian experience in hosting large numbers of non-Estonian military forces is long; Peter the Great established the first Russian military bases in the Estonia in the latter part of the seventeenth-century. During World War II, Germany seized many of the bases for their own use, until the Soviets retook the region and maintained a significant military presence in Estonia.

As a result of the co-operation between the Ministry of Defence and Ministry of Environment, 1,565 military facilities, including 14 major air bases, 3 military ports and 3 training areas of the armed forces of the former Soviet Union were identified on 87,000 hectares in Estonian Republic, covering 2 % of the whole country. The most militarised county was Harju County, where were nearly half of the military sites were located (546 facilities and 48,040 hectares respectively). The most militarized city was Tallinn, the capital of Estonia, with its 291 military sites.

In 1992 Estonian Ministry of Environment initiated the project "Assessment and liquidation of damage caused to the environment by former Soviet Union Army", which was carried out until 1995. The inventory of former Soviet Union military sites was carried out following the methodology introduced through joint project with German Federal Ministry for the Environment, the Nature Conservation and Nuclear Safety. As a result of the investigations, it turned out that 20 sites were extremely polluted, the majority of which were the air bases, big fuel storages, harbours, and parts of the rocket bases. One hundred fifty-five (155) sites had average pollution and were littered with dangerous substances. They included harbours, fuel storages, majority of the rocket bases, etc. Two hundred eighty (280) sites had minor pollution and approximately 300 sites were less dangerous and less polluted. More than half of the armed forces facilities of the former Soviet Union were polluted to some degree.

As a result of the investigations the environmental damage caused to the Estonia's environment is valued at 3.6 million (USA- billion) Euros. The possibility of getting compensation from the state that caused it is close to zero.

THE MAIN TASKS ON ENVIRONMENTAL PROTECTION

The Estonian Ministry of Defence uses approximately 90 facilities of the former Soviet Union military, which it is 7% of all amount former Soviet military facilities, covering approximately on 18,000 hectares. The large number of facilities is needed to provide the units of the Defence League located all over Estonia with necessary buildings and areas for carrying out training. The large part of the territory covers by Central Training Area (approximately half of the territory).

Because units of the Defence Ministry and the Defence Forces were forced to start using the facilities of the former armed forces of the former Soviet Union without cleaning up the existing pollution, attention needed to be paid to identify, localize and avoid the pollution. Taking into consideration little experience of the Defence Ministry and shortage of environmental specialists, also the lack of infrastructure, the work is carried out in cooperation with the Environmental Ministry and cooperation with other NATO and Partnership for Peace countries. At the present time there are joint, on-going projects with several countries (USA, Germany, Sweden, and Finland).

The objects, divided according to the main pollution, are as follows:

1. Ammunition factory: (The state joint-stock company E-ARSENAL) chemically polluted waters from galvanic and other production lines containing heavy metals and cyanides, the soil, and water polluted with oil products, emissions of toxic substances while painting, and solid wastes.
2. Fuel storages of oil products: soil and ground water contain oil products.
3. Naval Base Mine Harbour: soil and ground water polluted with oil products, sewage from ships, and chemical on the shore and in the waters, dangerous wastes and substances.
4. Ämari Air bases: soil and water polluted with fuel.
5. Training Area: soil erosion, unexploded ammunition, abandoned and unusable equipment and targets, dangerous wastes and substances.
6. Garrisons: polluted soil and ground water and soil from the buildings.

The main environmental task is to identify, localise and later on remediate the environmental pollution on military facilities of the former Soviet Union used by Defence Ministry and Defence Forces and the prevention of further pollution. Today we have conducted environmental investigation at 90% of Defence Forces facilities and determined the pollution areas. Biggest problems we have in the Ämari Airfield and in the Naval Base Mine harbour.

PLANS FOR 2003 AND MEDIUM TERM GOALS

Estonian Government has Membership Action Plan for NATO Membership. Every year, since 1999 Estonia publishes the Annual National Programme. In the programme Chapter 2.2.9 is Environmental Protection. Estonian Defence Forces (EDF) has for its 2003 goals:

- To complete the environmental study of the Central Training Area Complex
- To continue the environmental study of the Naval Base
- To continue the Ämari Air Force Base environmental project
- To continue the development of environmental regulations for EDF
- To raise environmental awareness within EDF personnel and to develop EDF educational programmes in environmental issues

Estonian Defence Forces have further medium term goals:

- To raise environmental awareness among the Estonian Defence Forces personnel and to develop respective educational programs for Estonian Defence Forces personnel
- To conduct a study in cooperation with the Ministry of Environment on the conditions and the level of pollution in the military installations, in order to prevent the future contamination of military installations
- To conduct an ecological study of military installations that have potentially high environmental risk

CURRENT SITUATION AND IMPLEMENTATION OF PLANS

The environmental protection activities have two main goals:

1. To carry out the clean-up of the former soviet military bases in cooperation with civilian organizations, foreign companies and partners.
2. To increase environmental awareness and implement the principles of environmental protection in the Estonian Defence Forces.

The Estonian Defence Forces intends to localize the above-mentioned environmental pollution. In order avoid further pollution it is necessary to apply the following measures:

1. To develop extensive cooperation with the Environmental Ministry of the Republic and other NATO and Partnership for Peace country, ministries of defence, and environmental ministries considering their ample experience in remediating and anticipating environmental damages caused by military units.

2. While planning the environmental undertakings and aims, the defence ministry has to follow the national environmental policy, laws and norms.
3. To determine level of use at the present moment and in the future, in order to develop the infrastructure to prevent environmental pollution (sewerage, oil gatherers, waste depositories, etc.).
4. In developing the defence forces, proceed from the principals of rational development of the nature and its resources (The Law of Economical Development).
5. Cleanup of pollution and environmental damage caused by military training where special equipment is required and reduce their use and reduce or prevent damage to fauna and flora. Minimising noise and air pollution in the air bases and during the training's.
6. Cleaning garrisons of abandoned garbage and making storage sites for garbage, in order to avoid further pollution.
7. Environmentally safe use and storage of materials and equipment used by the defence forces.
8. Examine the mechanical state of motor transport means, ships and aeronautical engineering and their correspondence to the established standards (oil leakage's, petrol leakage, the correspondence of exhaust gases to the norms).
9. The safe handling and storage of chemical wastes created during training of the defence forces.
10. The removal of unexploded ammunition from the polygons after the military training.
11. Securing the electromagnetic radiation flow from the radar stations while using the radar and in radio stations to a safe level to the population and nature.

Today the Ministry of Defence and the Defence Forces have carried out the following major pollution reduction activities:

- In cooperation with a Norwegian private company, the Paldiski North Harbour (4 vessels) and the Tallinn Mine Harbour (13 vessels) have been cleared of sunken ships of the armed forces of the former Soviet Union. The wrecks of ships littered most of the pier line and oil was constantly leaking into water from these vessels.
- In co-operation with the Swedish Royal Navy, German Bundesmarine, the Latvian Navy, and the Lithuanian Navy, the Estonian Navy has cleaned the waters and fairway's of the unexploded ammunition and other explosives.
- The engineer service of the Defence Forces has destroyed explosives left by the armed forces of the Russian Federation and Second World War on the island of Naissaare, in Astangu

explosive storages, on the Paldiski peninsula and Utsali Training Area. The activity is on-going over the several years.

- According to the development plan of the fuel economy, the majority of fuel tanks in the bigger military units have been substituted with up-to-date container tanks.
- In co-operation with USA, MoD, EPA and Finnish Armed Forces new Central Training Area for Estonian Defence Forces was established and completion of Environmental Management and Safety Plan.
- Reconstruction plan is being worked out for the Tallinn Mine Harbour of the Estonian Naval Base and Ämari Air base of the Estonian Air Force Base with assistance of US and Swedish military Experts.

ENVIRONMENTAL TRAINING

The most important task is to raise the environmental awareness of the conscript, NCO and Officers. The states belonging to NATO and Partnership for Peace countries from Scandinavia have developed a solid environmental training system. Many Estonian Officers and NCOs participate in different foreign Environmental Protection courses. Estonian Defence Forces work with Environmental Training, under leading Swedish and US colleges and with Latvian and Lithuanian Defence Forces.

In 2002, the handbook "Environmental Protection in Defence Forces" was published. This book is a adaptation of Swedish handbook "Handbok miljö för Försvarsmarkten (H Miljö)" for Estonian laws and Estonian Defence Forces structure. In 2003, CD-ROM with Environmental Protection training movies for conscripts is to be released. Main elements of environmental education are environmental awareness, specific environmental procedures and measures, and supervisory responsibilities.

Environmental awareness consist environmental protection (water resource protection, air and atmosphere quality protection, vegetation and soil protection (including forests), noise abatement, landscape quality protection, natural resources and habitat protection, material and waste management and pollution prevention), resource conservation (heritage protection (natural and man-made) and resource protection in context of sustainable development (energy conservation, reduction in the use of non-renewable resources and waste minimisation and recycling) and environmental policy (domestic environmental Policy, Host Nation's Environmental Policy, NATO and EU environmental policy and multilateral environmental agreements.

Specific environmental procedures relating to procedures or measures of a specific nature are designed to ensure environmental protection during the performance of duty need more detailed environmental education, probably given to smaller groups of personnel. For example, engineer construction, aircraft and vehicle maintenance, waste handling at sea and fuel handling. Supervisory responsibilities are an area of particular concern to commanders and their nominated supervisory staff. The need for professional environmental expertise may require ready access to specialised experts/advisors. Particular attention will have to be given to environmental policy and guidance, risk management and planning.

Environmental education should be incorporated into existing training program. Opportunities include:

- Individual training: the basic training courses, Junior Officer courses, NCO courses - majority of environmental awareness, procedures and supervisory material
- Collective training: This provides the opportunity for initiatives such as the presentation of standing orders for exercises and training areas, waste handling and pollution prevention
- Continuation training: Environmental knowledge and training will need to be continuously updated and refresher training provided as legislation and working practices develop.

In Estonian Defence Forces have determined the Environmental Protection skills and qualification demands for different command levels:

- Conscript (Soldiers, Junior NCO-s)
- COY, Plt. Commanders, Senior NCO-s
- Specialist officers and NCO-s
- Base, Unit Commanders, Staff officers
- General Staff, A/N/AF Staff officers, Civil servants in MOD
- Minister of Defence, Supreme Commander, CoS, CaA, CoN, CoAF
- Environmental officer

Routine training of the Environmental Protection in the Estonian Defence Forces:

- Conscripts training:
  - Soldier's Basic course - 1 hour
  - Soldier's Speciality course - 1 hour
  - Training on section and platoon level – environmental protection instructions and on-job training during exercises
  - Junior NCO course - environmental protection instructions and on-job training during exercises
- NCO School (Senior NCO training) - environmental protection instructions and on-job training during exercises
- Estonian National Defence College (officer training)

- Environmental protection course (40 hours) for Logistical and Border Guard officers (since 2001)
- Environmental protection instructions and on-job training during exercises for all specialities

Plans for future are in the environmental training follows:

- Negotiations with Bundeswehr over the future co-operation in NBC-Defence and EP training in 2002
- Including the Environmental Protection course into the Senior NCO curricula in NCO Academy in 2003
- Starting of the Environmental officers course in co-operation with Sweden, USA, Latvia and Lithuania in 2003
- Participation in NATO Environmental Protection training workgroup in 2003

The Estonian Defence Forces are real defensive forces when definition "defence" includes nature.

NATO Environmental Doctrine from Hungarian Point of View

Zoltan Jaczo, Arpad Vincze
Miklos Zrinyi National Defence University, Department of NBC Defence and Environmental Security, Budapest, Hungary

INTRODUCTION

The publication of the joint NATO doctrine for the environmental protection of NATO led operations and exercises, together with the statement of the commanders' responsibility and education policy for the preventive protection of the environment, was a very important event for the experts in the field of environmental protection in the Hungarian military sector.

The joint NATO doctrine for environmental protection was elaborated by the EPWG [1] as a result of more then four years of work. The complex nature of a general environmental protection policy applicable for military operations and exercises caused continuous conceptial changes during the elaboration process. The process had three phases. In the first phase the very detailed environmental demands were collectes and formulated. In the second phase, these detailed environmental demands were "softened" and made generally applicable. This contradictive situation had been resolved in the third phase by the implementation of the modern approach of environmental management systems into the doctrine.

Due to this change of conception, - in contrast to the original plans -, the draft formulation did not contain a general environmental policy of NATO after 2001 anymore. Since there is a great variation in the civil environmental protection regulation among member states, the doctrine was published as STANAG 7141 entitled "JOINT NATO DOCTRINE FOR ENVIRONMENTAL PROTECTION DURING NATO LED OPERATIONS AND EXERCISES".

Representatives of the Hungarian Ministry of Defence (HMD) – before its promulgation – had officially declared its ratification without reservation and implementation in the Hungarian Defence Forces (HDF) within one year after promulgation [2].

What does the implementation mean? According to the 3rd point of the agreement, "this STANAG is implemented when a nation has issued the necessary orders/instructions to the forces concerned putting the

K. Mahutova et al. (eds.), Defense and the Environment: Effective Scientific Communication, 43–50.

procedures detailed in this agreement into effect". Ratification without reservation means that the necessary orders/instructions must be formulated and put into effect in all area of the agreement before November 2003.

This article is intended to express the authors' view of the implementation process in Hungary, highlighting the most problematic areas.

Statements of this article only represents the authors' personal view, and therefore do not represent that of the Hungarian Ministry of Defence nor of the Hungarian Republic.

STRUCTURE OF THE AGREEMENT

The agreement part of the document is short and complies with the directions of AAP-3(H) [3]. The aim of the agreement declared in the first point of the standardization is to "to state NATO environmental doctrine for NATO led operations and exercises and to provide guidance in environmental planning for all military activities." The second part lists the ANNEXES to be agreed with:

ANNEX A: the NATO environmental doctrine;
ANNEX B: the commander's responsibilities;
ANNEX C: the environmental education guidelines.

The third point of the agreement – as quoted above – states the condition of implementation.

In addition to the ANNEXES, the document provides the list of related documents, which contains references and/or regulations that should be considered during the application of this STANAG. One of the related documents is connected to logistics, eight out of them are related to the navy, one deal with insect and rodent control and one deals with field sanitary requirements. From an environmental protection point of view, the most important documents are the NATO CCMS Reports 240 and 240B [4,5], which summarizes the findings of the three years project "Environmental Management Systems in the Military Sector". Not less relevant is that the ISO 14000 series referring to environmental management systems can be found among the related documents [6,7].

It is very crucial to realize that the inclusion of the modern environmental approach into the doctrine, which is based on the concept of prevention, identification, minimizing and manage of environmental impacts is of utmost importance. The greatest advantage of this regulation is that it's requirements and methods of problem solving are based on the most recent achievements of environmental sciences:

- System based preventive environmental protection,
- Theory of sustainable development and
- Need for the rise of environmental awareness.

ENVIRONMENTAL DOCTRINE

In the introduction (1st point) it is stated, „While meeting their military mission, NATO Forces should be committed to taking all reasonably achievable measures to protect the environment. To achieve this, commanders must know how NATO led operations and exercises affect and are affected by the environment. Environmental planning is an essential process to ensuring appropriate environmental protection".

The fact that STANAG 7141 uses exactly the same definition for environment and environmental impact as appears in the ISO 14001 international standard, illustrates the integration of the standardization process between the civil and military regulations. According to the introduction, the protection of the environment becomes the duty of the armed forces within reasonable bounds. To understand this, one should consider the environmental planning guidelines in the six point of this annex, which makes it clear that „the principal responsibility of all commanders lies in the achievement of their military tasks". The military tasks therefore have higher priority than environmental protection. It is also obvious that the priority of environmental protection is different in different tasks such as exercises, humanitarian missions and peacekeeping operations or in war.

The introduction also declares the principle of the identification of dual effects, taking one of the three base pillars of environmental management systems (identification, minimizing and handling of impacts). The duality lies in the requirement that the commanders in all level must know how military operations and exercises affect the elements of the environment (including potential threats) and must consider how far the environment affect the possible ways of achieving the military tasks.

It can be seen therefore, that the understanding of significant environmental impacts has an important role already in the planning stage of the mission. This is manifested in the accentuation of environmental planning, highlighting the preventive character of the whole environmental doctrine.

The first element of environmental planning is the identification of environmental problems, for which solutions must be found. In this process commanders can early identify significant environmental impacts that have great effect on the success of the mission. It is obvious therefore that in the elaboration and choice of possible strategies of the mission, the evaluation of the negative environmental effects shall play an important role.

For effective planning the knowledge of public regulations is also essential. Military environmental protection has always had problems to comply with civil environmental regulations. This was mainly due to the special situation of military operations as partly „out of law", which was primarily caused by the secrecy of these missions. The strengthening of civil control, the development of regulations and the growing interest of the society have made this situation indefensible.

In Hungary environmental regulation is based on the Act of Environment (1995) [8]. For this modern, EU conform law does not exclude the military sector, facilities of the Hungarian Defence Forces (except strategic ones) can be the subject of inspection by the public environmental protection authority at short notice. Furthermore, in the case of non-compliance with the law, the commander can be charged and, if fault is faund the commander can be fined.

The notion of environmental crime has emerged in Hungarian law [9]. Violant can be penalized with imprisonment. The military command system is based on the unconditional execution of commands, which can only be denied if the execution would result in the commitment of crime. It is extremely important therefore for the commanders to know and appropriately use environmental law [10].

Obviously – like other civil companies – large military facilities cannot fulfill all environmental rules, therefore early identification of great environmental impacts, knowledge of priorities and the thorough planning of the reduction of negative effects are essential.

The third point of ANNEX A states "environmental damage may be an inevitable consequence of operations". Therefore "environmental planning should minimize these effects without compromising either operational or training requirements".

According to point 4, "commanders should ensure environmental risk management is integrated into the overall planning for exercises and operations", where environmental risk management is defined as "the process of detecting, assessing and controlling risk arising from operational factors together with balancing risk with mission benefits".

This idea is on one hand the implementation of the ISO 14001 standard, which requires the identification, minimization and management of significant environmental impacts. On the other hand, the compromise nature of ISO 14001 is well integrated with the priority of mission accomplishment, balancing thereby between environmental (and not exceptionally) risks and strategic advantages due to the successful completing of mission. In other words, the cost of environmental damage and remediation should be in proportion with the importance of mission accomplishment.

The doctrine require that "the risks associated with efforts to protect the environment will be different for each phase and should therefore be considered separately prior to, during and after the mission or exercise". It is important to note that risk assessment involves not only a priori events, but also the likelihood of environmental contingencies, such as accidents. The accidental damage to the equipment caused by the attack of the enemy should also be considered.

Since environmental planning is the basic idea of the doctrine, separate guidelines are provided for NATO led operations and exercises. To be able to comply with the doctrine, commanders should ensure that "their Planning staffs are properly trained on environmental aspects". This requirement makes the involvement of environmental experts into the overall planning for exercises and operations.

The key elements of environmental risk management framework are formulated as follows:

a.) Commander's environmental policy or guidance.
b.) Environmental planning.
c.) Implementation.
d.) Checking and corrective actions.
e.) After action review.

These elements of environmental risk management are very similar to the system-based ISO 14001 standard for the implementation and usage of environmental management systems [6]. The basic idea of the latter is the periodic repeat of the "*plan→do→check→act*" cycle, with the formulation of new goals in the planning phase and with the implementation of these the assurance of the continuous enhancement of the results (i.e. environmental achievements) as illustrated in Figure 1.

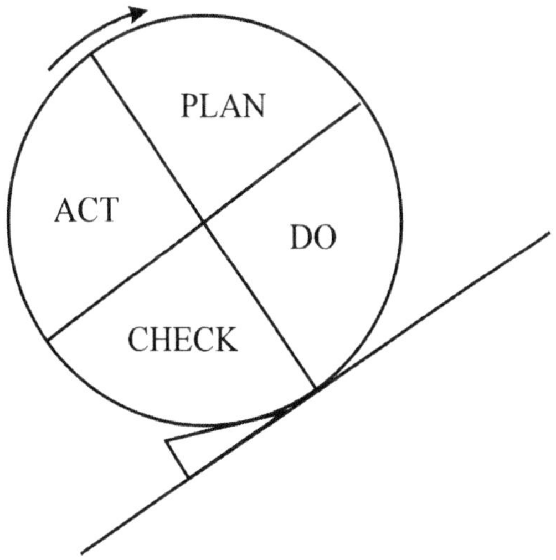

Figure 1. The "plan→do→check→act" cycle of environmental management systems as used in ISO 14001 standard.

The framework of the ISO 14001 type of cycle is created by a clear and determinate environmental policy. For the environmental policy the ISO 14001 standard formulates many contextual requirements, out of which the most important are that environmental policy must:

- comply with the operation of the given body and its environmental impacts,
- contain commitment for the continuous improvement and for the prevention of environmental pollution
- contain commitment for the fulfillment of environmental laws,
- provide a framework for the formulation of environmental goals and their inspection,
- be well documented and introduced to the staff,
- be accessible to the public.

As can be seen, commander's environmental policy could comply with most of these requirements.

In the original concept of STANAG 7141 the issue of the „Environmental policy of NATO allied forces" as a separate ANNEX was considered. It was a good decision however, that in the final edition the formulation of the environmental policy remained in the commanders' responsibility making it possible to best fit to the actual circumstances.

It is easy to realize, that the risk management philosophy of STANAG 7141 is based on elements of the ISO 14001 cycle. However continual improvement can only be achieved stepwise as illustrated in Figure 2.

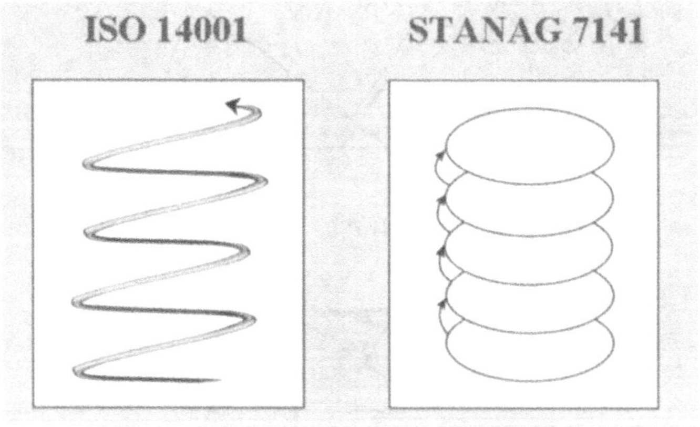

Figure 2. The dynamics of continual improvement in ISO 14001 and STANAG 7141, respectively.

The stepwise improvement is achieved by applying the experimental finding of the previous mission to the planning phase of next mission (i.e. next cycle). This way the continual improvement of environmental planning and implementation can be guaranteed. This can only work however, if the following conditions are met:

- "Environmental" officers with adequate experience are involved into the planning process,
- Experiments are provided to all officers responsible for planning.

It should also be noted, that – in contrast to EMAS II standard used in the German and English industry – ISO 14001 does not demand overall fulfillment, but requires the commitment for the continual improvement and for the prevention of environmental pollution. This is very important for the military sector, since this is the most efficient way to achieve the desired conditions. This is well manifested in the formulation of STANAG 7141 and NATO CCMS Report 240.

IMPLEMENTATION IN HUNGARY

STANAG 7141 is to be implemented in Hungary in November of 2003. The fulfillment of the recommendations of the doctrine is only achieved, if the high command of the Hungarian Defence Forces publishes the necessary internal regulations, which permit the principles laid down in the environmental doctrine to be enforced and ensure commanders' ability for execution of their duties. Additionally, these regulations would facilitate the introduction of integrated environmental education. These internal regulations are yet to be published.

We believe that the proper training of military environmental experts, keeping them in the system and provide further training for them are extremely important. As far as training and education are concerned, all the necessary infrastructure and personal conditions are at the disposal of the Hungarian Defence Forces via the National Defence University and other public universities in Hungary. Training of such environmental experts has already begun.

The major difficulty lies in the fact, that Hungarian Defence Forces facing very radical reforms. One of the main tasks is that HDF must be transformed to fully professional armed force by 2006. This has the immediate consequence of cutting back the number of soldiers, which makes difficult the introduction of military environmental experts into the system.

Rearmament of the Hungarian Armed Forces expected to take place within 2010 - 2015. Environmental protection in the Hungarian military sector in the next decade therefore will depend on the success of modern environmental approach together with the old, inherited techniques.

It is expected therefore, that the Hungarian Ministry of Defence will recommend the high command of the Hungarian Defence Forces to the work out the necessary internal regulations at first stage only for those forces that are participating in NATO led missions.

REFERENCES

1. Environmental Protection Working Group, EPWG
2. Statement 1161/3/2002 of the Under Secretary of State of the Hungarian Ministry of Defence. Defence Publications, 2002/23, p. 1547. (in Hungarian)
3. NATO Allied Administrative Publication AAP-3(H), March 2001, "Procedures for the Development, Preparation, Production and the Updating of NATO Standardization Agreements (STANAGs) and Allied Pubs (APs)".
4. Report Nr 240 – Environmental Management Systems in the Military Sector, Final Report, NATO CCMS, Brussels, 2000, pp. 115.
5. Complementary Guidelines for the NATO/CCMS Final Report nr. 240, (Workshop) Thun, Switzerland, 22-26 January 2001, pp. 121
6. ISO 14001:1996 – Environmental Management Systems. Specifications with guidance for use. pp. 41.
7. ISO 14004:1996 – Environmental Management Systems. General guidelines on principles, systems and supporting techniques. pp.75.
8. Act LIII of 1995 on the General Rules of Environmental Protection.
9. Act IV of 1978. 280. § (1) paragraph.
10. Z. Jaczo and J. Solymosi: "Some questions about environmental regulation in the Hungarian Defence Forces". ZMNE Publications, Budapest, 2001, p. 138-144. (in Hungarian)

STANAG 7141 EP – Plan of Implementation in Poland

Stanislaw Wilczkowiak
Department of Defence Affairs, Ministry of Environment, Warsaw, Poland

BACKGROUND

As concerns grow for maintaining and improving the quality of the environment and protecting human health, NATO as well NATO member States are increasingly turning their attention to the potential environmental impacts of their activities. The achievement of sound environmental performance requires organizational commitment to a systematic approach and continual improvement of the environmental management system. Edited on 23rd May, 2001 NATO Standardization Agreement "Joint NATO Doctrine for Environmental Protection During NATO Led Operations and Exercises" (STANAG 7141EP) provides assistance to participating nations in environmental planning for all military activities. Poland signed STANAG 7141EP on July 17, 2002 and declared to be ready for full implementation by April 2004.

We are considering three types of NATO led operations and exercises:

1. on the territory of Poland;
2. on the territories of other NATO member states;
3. outside the territories of NATO member states.

When we host other military forces on our territory it is necessary to deliver to NATO commanders, proper information about our environmental legislation, state of the environment, environmental services etc. needed for environmental planning and risk management. When our military forces are hosted by other NATO member states, we can expect that the host country should play a key role in the process of environmental planning and environmental risk management. When NATO mission or exercises take place outside the territories of NATO member states, we have to be prepared for the potential lack of environmental information, weak performance of environmental institution and services in these places. In this case there is a need for a special preparation and self-sufficiency.

K. Mahutova et al. (eds.), Defense and the Environment: Effective Scientific Communication, 51–61.

In this paper I would like to focus on case of NATO led operations and exercises in Poland.

There is no doubt that the process of implementation of STANAG depend on numerous factors, such as existence of:

- national environmental policy;
- legal regulations concerning environmental protection;
- state of natural environment;
- environmental institutional, socioeconomic and technological capacity;
- military techniques, structures and types of activities;
- civilian, military and international cooperation.

NATIONAL ENVIRONMENTAL POLICY.

In 1991, the Sejm (parliament) of the Republic of Poland adopted the first National Environmental Policy, which sets goals and actions for improving the state of the environment. In this document Poland recognizes sustainable development as the very basis of its social and economic progress. The principle of sustainable development has also been confirmed by Article 5 of the new Constitution of the Republic of Poland (1997):

> "The Republic of Poland protects its independence and inviolability of its territory, ensures human and citizens liberties and rights as well as its citizens' safety, stands on guard of its national heritage and ensures environmental protection in accordance with the sustainable development principle".

The Ministry of the Environment prepared the second National Environmental Policy in 2000. This document provides environmental policy principles, objectives, tools and instruments for the next 10 years. The following principles have a very important impact on the Polish defense sector:

- the principle of sustainable development;
- the precautionary principle;
- the principle of integration of environmental policy into sector policies;
- the "polluter pays" principle;
- the principle of prevention; and
- the principle of applying the best available technique.

The National Environmental Policy recognizes environmentally sound management as a very important tool for the regular identification, and of further consequence, the implementation of environmental objectives leading to limitation of potential environmental impacts and use of environmental resources. The main objective of the Second National Environmental Policy is to provide the environmental safety, social infrastructure and natural resources for the population. It requires sound strategies and policies in particular sectors, as well as the management, compatible with the sustainable development strategy.

The Armed Forces are an important element within Poland's socio-political and economic system. Their size, equipment and area of land occupied require that they have to take all reasonably achievable measures to protect environment during their activities and exercises. In accordance with National Environmental Policy of Poland, the armed forces are subject to all normative environmental regulations and policies. Moreover, environmental liability applies both to military organizations and to personal liability of the commanders at all levels of command. The implementation of STANAG 7141EP in Military Sector is in line with the National Environmental Policy.

LEGAL REGULATIONS CONCERNING ENVIRONMENTAL PROTECTION

Effective implementation of environmental policy requires a cohesive system of environmental legislation. Environmental law has a long tradition in Poland. After regaining independence in 1918, the young state immediately began its work on legal regulations in this area. The first Water Act was adopted in 1922, while the Nature Conservation Act of 1934 set the trend for nature conservation. In its day it was one of the most forward-thinking and modern Acts of Parliament of the kind anywhere in the world. In the last decade, Polish environmental legislation has been deeply restructured. We are very close to achieving following:

- compliance with Constitution;
- compliance with international commitments of Poland (especially EU);
- social acceptance;
- feasibility;
- environmental effectiveness;
- economic efficiency;

Currently in Poland, the protection of the environment is a constitutional obligation falling upon all citizens and authorities. Moreover, the Polish legal environmental protection system broadly conceived, comprises more than 120 legal instruments (Acts and the executive regulations for them). Among these, the most important are, as follows:

- the Spatial Planning Act of July 7th, 1994
- the Environmental Protection Law Act of April 27th, 2001
- the Water Law Act of July 18th,2001
- the Act of June 7th, 2001 on the collective supply of water and removal of wastewaters
- the Geological and Mining Law Act of February 4th, 1994
- the Wastes Act of April 27th, 2001
- the Packaging and Packaging Wastes Act of May 11th, 2001
- the Agricultural and Forest Land Protection Act of February 3rd, 1995
- the Nature Conservation Act of October 16th, 1991
- the Forests Act of September 28th, 1991

THE NATURAL ENVIRONMENT IN POLAND

In order to effectively integrate environmental consideration for NATO- led exercises or operations, commanders should identify related characteristics of the environment.

Poland is a relatively large country, 312,700 km2 with population 38.6 million. Poland is situated in the temperate zone, with a climate transitional between the Western European Oceanic and the Eastern European Continental. Mean annual temperatures vary from 3°C in the mountains of the south to 8°C in the west-central part of the country. Mean yearly precipitation totals are in the range 550-1,300 mm with the respective extremes in the central part of the country and in the mountains.

There are a wide variety of types of topography, with the lowlands occupy 91 % of the country and the uplands only 8% (with true mountains accounting for only about 1 % of Poland). The present landscape is almost entirely anthropogenic in character, with changes having been wrought by agriculture, forestry and development (of settlement, industry and transport), leading to habitat fragmentation, and the creation of barriers isolating species populations. New types of ecosystem have also been created, including those of strip mining, spoil heaps and reservoirs. The intensity of human pressure depends on patterns of land use and reaches a peak in the industrial agglomerations of southern and central Poland.

Despite the changes that human pressure brings upon the environment, Poland is one of the European countries with the highest levels of biological diversity. This is linked with favorable natural conditions and variable human impacts that, in spite of local problems, are generally of lesser intensity than in other European countries. The wealth of flora and fauna in some areas is unique, not only for Europe, but in the whole world. In order to protect its natural treasures, Poland has created a system of protected areas (National Parks, Nature Reserves, Landscape Parks, Areas of Protected Landscape, etc.), which now extends over more than 26% of the country and there are plans to bring additional valuable areas under protection. The ecological corridors of ECONET link extensive and precious ecosystems of national or international significance.

ENVIRONMENTAL INSTITUTIONAL, SOCIOECONOMIC AND TECHNOLOGICAL CAPABILITY

Poland has already developed efficient organizational structures supporting the national environmental policy, in particular in the fields:

- financing for environmental protection (National Fund for Environmental Protection and Water Management) and a system of provincial and local community funds, Eco Fund and other public and private foundations;
- environmental supervision system (Environmental Inspection with a network of labs);

Recently, a number of agencies were established as public or joint public- private structures acting partly or entirely in environmental protection sector.

It is expected that institutional capability should be reinforced in following areas:

- collection and dissemination information for decision making (information on BAT and "cleaner production", state of the environment, GIS);
- approximation to EU environmental requirements.

Poland's education system has long acted to shape environmental public awareness and directed itself to issues under the heading of sustainable development. Environmental issues are integrated within all levels of education. Studies in environmental protection are provided at more than 10 universities. Funds from the National Fund for Environmental Protection and Water Management support actions and campaigns of environmental content, including publishing books and journals and the broadcasting of television and radio programmers to a wide audience.

Economic and technological environmental capability of Poland has steadily increased. The introduction to the economy of environment friendly production technologies is very fast in all sectors. The investment in production, especially those of significant influence on their surroundings, has become subject to a requirement to conduct an

Environmental Impact Assessment. The transfer of environmental friendly technologies is one of the most important elements in multilateral and bilateral cooperation of a technical and scientific nature. Full implementation of BAT principle and IPPC procedure will be very important step for improving economic and technological capacity of environmental. We can see a very fast development of environmental services.

IMPLEMENTATION OF STANAG 7141EP

Poland has been a member of NATO since March 1999. We are active in the work of all of the bodies of the alliance. Poland military forces took part in a number of NATO led operations and exercises in the Poland, other NATO member states or within NATO territories (for example: SFOR, KFOR-Kosovo, exercises "Strong Resolve"). We are well experienced in resolving environmental issues during multi-national military exercises on Polish soil.

General environmental issues defined in STANAG 7141 EP are:

- environmental planning;
- environmental impacts on training areas;
- environmental risk management; and
- environmental training and education.

In accordance with STANAG 7141EP requirements, the following tasks should be in framework of environmental planning:

- identification of operational activities that could have a potential impact on the environment;
- identification of alternatives and contingences;
- identification of characteristics of the environment that may be affected by or have an impact on military activities;
- identification of potential effects caused by the military activities, including the effects of alternatives and contingencies;
- identification of feasible mitigation measures to reduce the risk to the environment and human health and safety;
- measures for pollution prevention and resources conservation;
- clean up and remediation;
- determination what national and/or international environmental laws are applicable to the operation or exercise; and
- determination of limits and restrictions imposed by applicable environmental regulations or policy.

In order to fulfill these tasks it is necessary to have access to proper information, methods and tools. The most sophisticated tools, for example expert systems, Geographic Information Systems, are too costly to implement on lower level of command. Still very effective are such methods as analogs, checklists, expert opinion, and matrices.

During peacetime, special attention should be taken for environmental protection on military training areas. There is an obligation to:

- establish environmental baselines for sites;
- ensure protection of nature, flora and fauna;
- ensure protection of landscapes, historical monuments and watercourses;
- carry out periodic inspection of sites and monitor and make changes;
- ensure that necessary repair works have been carried out;
- incorporate any lessons learned in future planning; and
- work with local authorities and communities to identify and resolve problems.

These tasks usually belong to the commanders of military training areas. In recent years we have noticed, in NATO countries, a number of new ideas and tools of environmental management in military areas.

STANAG 7141 EP defines following key elements of environmental risk management:

- clear guidance on environmental protection for the exercise or operation;
- environmental planning and environmental contingences;
- clear assignment of responsibilities and resources;
- environmental training;
- monitoring of activities; and
- post-action review.

The responsibility for environmental risk management principally lies on commanders. They should demonstrate leadership and awareness in environmental protection, identify and assign clear responsibilities and resources, as well as, ensuring compliance.

STANAG 7141 EP gives also clear guidelines for environmental training and education. As a principle it assumes that environmental training should be incorporated within existing training programs in pragmatic manner.

The key factors of leading successful implementation of NATO STANAG are as follows:

- proper allocation of resources;
- assignments of responsibilities; and
- on-going evaluations of practice, procedures and processes.

We began our analysis from the evaluation of existing responsibilities, practices, procedures and processes of environmental protection in our military sector. In 2002 a special working group from the Ministry of Defense analyzed following issues:

- the basic regulation and other documents;
- the organization structure of environmental protection in military sector in Poland;
- typical exercises of Polish military forces and their impact on environment;
- typical joint operation and exercises and their impact on environmental;
- typical hazardous materials and substances used in military and their impact on environment;
- assessments of needed resources and changes in organization for environmental protection in Polish military sector;

The output of the work was a document titled "Principles and Procedures of Implementation of STANAG 7141EP", which includes results of above analysis and the following:

- assumptions for amendments of Minister's of National Defense environmental regulation;
- the basic need for development of environmental information infrastructure;
- the need for creation of environmental plans of training areas;
- assumptions for environmental guidelines for commanders;
- assessment of additional cost of STANAG 7141EP implementation;

This document was prepared in consultation with other ministers (especially Ministry of the Environment).

On the base of this analysis, a new regulation on environmental protection at the military sector – Decision No 57/MoD (entry into force - December 2002) was prepared. The regulation assigns tasks and responsibilities to the organizational units of military sector. According to this regulation, the Ministry of National Defense is responsible for preparation of environmental policy of military sector, programs of environmental protection, coordination of environmental education and training, budgeting, environmental auditing, assessing the state of the of the environment in military sector and compliance, coordination of radiation detection system in the Armed Forces, Central Database on Dangerous Chemical Substances, research and development, organization for recovery of polluted areas, international cooperation and collaboration with other institutions.

A key roles in environmental protection belongs to following bodies in the Ministry of National Defense :

- Secretary of State (MoND);
- The Chief of the General Staff of the Polish Army
- Department of Infrastructure (where is embedded Division of Environmental Protection Branch);
- Armament Policy Department;
- Department of Education and Promotion of Defense
- Press and Information Office

The commanders of the Armed Forces, Air Defense Forces and Navy are responsible for implementation of Defense Environmental Policy, planning and supervision of environmental activities in their areas of responsibilities. Other military commanders should follow existing guidelines, programs and should also ensure compliance with environmental regulations.

The next steps planned in the process of implementation of STANAG 7141EP are as follows:

- preparation of necessary orders/instructions for Commanders of the units of Polish Armed Forces, The Air and Air Defense Forces and the Navy;
- preparation of guidelines for environmental protection in the military sector and specific guidelines to Armed Forces, Air and Air Defense Forces and Navy;
- elaboration of plans for environmental protection for military training areas; and
- development of an environmental information system.

The additional cost of implementation of STANAG 7141EP has been estimated on the level of approximately $2 million USD.

In the process of implementation of STANAG 7141EP we can identify good opportunities, as well as, some problems.

THE IDENTIFIED OPPORTUNITIES

1. Strong support from the Ministry of the Environment:

In 1997, the Ministers of National Defense and the Environment signed an agreement on cooperation in the field of environmental protection. In particular, the objectives of agreement included environmental education and training courses. This initiative has fostered the development of large number of interesting forms of cooperation including, among others:

- organization of annual competitions between diploma dissertations of higher military schools, military academies and post-graduate studies at those schools, entitled “The army protects the natural environment”;
- organization of readers’ competitions concerning environmental protection, addressed to conscripts;
- joint financing and publication environmental articles in popular military periodicals;
- joint organization of seminars and scientific conferences
- publication the guidelines for the Hosted Armed Forces “How and why should we protect the natural environment” by Ministry of the Environment.

The cooperation between these institutions ensures better implementation and quality of projects and their effectiveness in achieving the common goals of the Ministries of the Environment and National Defense.

2. Ability to finance some environmental projects from the National Fund for Environmental Protection and Water Management.

3. Active cooperation in framework of the NATO Committee for the Challenges of Modern Society. Poland leads the Pilot Study on “Forms of environmental education in the armed forces and their impact on creation of pro-environmental attitudes”.

4. Easy access to and good quality of civilian environmental services.

5. Lessons learned form previous NATO led operations and exercises.

THE IDENTIFIED PROBLEMS

1. Low budget of the Ministry of National Defense;
2. Poor availability and quality of environmental information, particularly on the local level.
3. Lack of modern tools for environmental impact assessment in the Ministry of National Defense.

CONCLUSIONS

1. The development of capability for environmental planning and environmental risk management during military operations and

exercises should be on-going process, integrated with the development of military as a whole.

2. The decision support system for environmental protection in the military sector should be easy to use and sufficiently flexible to analyze a variety of impact scenarios.

3. Well organized international cooperation between militaries in resolving environmental issues, as well cooperation between militaries and civilian environmental society, have a crucial role in time of high expectations and budget constraints.

REFERENCES

1. NATO, STANAG 7141EP, edition1, 2001;
2. Ministry of the Environment, How and why should we protect the natural environment- Guidelines for hosted armed forces.2002 from:
3. http://mos.gov.pl/mos/publicac/index-eng.htm
4. MoD. Decision No57 of MoD; Official Journal of MoD. 2002.

PART II:

ENVIRONMENTAL MANAGEMENT SYSTEMS AND PUBLIC PARTICIPATION

Public Participation in Decisions Relating to the Environmental Management of Ministry of Defence Sites

Simon M. Clark
Directorate of Safety, Environment and Fire Policy, UK Ministry of Defence, London, United Kingdom

INTRODUCTION

Traditionally, defence based organisations operate on a hierarchical command structure with decision making carried out at high levels. The decision-making is independent of external stakeholders and takes no account of factors external to purely military considerations. This is, and will remain, the best and most efficient way to manage military operations, particularly those of a sensitive nature or involving conflict.

However, the work of Defence Ministries is not confined to military operations. In many cases, Defence Ministries take on the stewardship of large tracts of land, primarily for training purposes, and the ownership of a large number of buildings, including "heritage sites". The ownership and stewardship on behalf of the nation as a whole brings with it the necessity for management of the environment, in the widest sense of the word. This management must be carried out in a way that allows military needs to be met but also preserves the environment for future use.

The participation of the public and of special interest groups in decision-making is a theme of the Government of the UK. Policy statements to this effect have been made in public, and the Cabinet Office (the part of the UK governmental structure with responsibilities for issues that affect many Departments of State) issued guidelines effective from the beginning of January 2001 regarding public participation. These are available on the Cabinet Office web site. (www.cabinet-office.gov.uk/servicefirst/2000/consult/code/consultationcode.htm) The UK Ministry of Defence (MOD) is bound by these guidelines and is therefore required to undertake public participation where practicable. Public participation would not required, for example, where it would compromise security or operational effectiveness.

With the constraint that public participation should not affect security or operations, it is clear that many functions within the MOD cannot be subject to such public involvement. One area where public participation can take place is in relation to the environmental management of the

K. Mahutova et al. (eds.), Defense and the Environment: Effective Scientific Communication, 65–70.

MOD estate. This paper outlines the ranges of public participation that could be undertaken and presents some of the UK MOD's experience with public participation on environmental matters.

PUBLIC PARTICIPATION

In considering public participation, it is important to understand what is meant. Public participation is now considered to cover a range of approaches to information arrangements that could involve the public. Commonly, public participation is given a series of categories and these are shown below in the order of increasing public involvement.

Information Giving. This is the traditional approach of telling the public what you have done after you have done it. There is no public involvement in any of the decisions that have been taken or the information that is supplied. It should be noted that the information given can be extensive - the key factor relates to the decision ability and not the information provided.
Consultation. Here, information is supplied to the public regarding what is planned for the future and their opinion is sought. There is an implication that the opinions supplied can change the decisions that will be taken, but there is no obligation to take the opinions received into account.
Bounded Dialogue. The organisation with the problem makes the decisions regarding strategy and overall direction of a project but then opens up the implementation to public participation. There is an undertaking that the implementation proposals made through the public participation process will be accepted by the decision maker. An example would be where it is decided that a building will be demolished, but the public would decide when and how this is done in order to minimise the impact on the local community.
Stakeholder Dialogue. All interested parties come together to resolve a particular problem or problem area. Progress can only be made when all parties involved in the dialogue agree on a particular point or way forward. This is known as reaching a consensus. The owner of holder of the problem involved has no special control over the direction the discussion takes and has undertaken to implement the consensus solution.

All of these approaches can be considered as public participation and the challenge in the Defence field is to select the correct public participation method for the problem identified. In the recent conflict in Iraq, public participation has been carried out by Information Giving, but this may not be appropriate where the problem involves the environmental management. In these situations, other forms of public participation may be more effective.

STAKEHOLDERS

Public participation makes frequent reference to "stakeholders", and it is important to understand what a stakeholder is and the different categories of stakeholder. Effectively, a stakeholder is any person or body that has an interest in the problem being discussed. Interest can range from being directly affected by the project to those who have an interest in the particular subject area.

Stakeholders can come as individuals or representing various groups - the MOD representative in any stakeholder dialogue represents both the MOD view and their own personal views. All that is required is to ensure that the rest of the stakeholders are clear about which view is being presented. Other parts of Government, the Regulators and Local Government officials are often stakeholders representing their own organisation's views.

The other groups that are present as stakeholders are the Non Governmental Organisations (NGOs) and the Community Based Organisations (CBOs). Greenpeace and Friends of the Earth are the best known of these organisations, but a wide variety of more local groups also exist and must be included as stakeholders if they so request. The most difficult stakeholder to identify and get to be involved in public participation is the one that is most required - the ordinary unaffiliated member of the public. In the absence of this key stakeholder, all other stakeholders should ensure that they are mindful of their secondary duty to the participation process as a member of the public.

STAKEHOLDER DIALOGUE

As already noted, stakeholder dialogue is the most open and involving method of public participation currently in use. It is for the problem holder to determine if stakeholder dialogue is appropriate and then to initiate that dialogue. It is very important that the limits of the dialogue are established at the outset, and that all parties understand those limits.

It must also be recognised that stakeholder dialogue is a slow process due to its requirement for consensus and in the final analysis may not deliver a useable outcome. Contingency plans should always be developed in case any public participation process fails.

The problem holder is not only responsible for initiating the stakeholder dialogue but is also required to finance the process. For many stakeholders, and particularly for members of NGOs, CBOs and the public, participation in stakeholder dialogue involves a significant commitment of time and effort. It is important to your dialogue to ensure that stakeholders become involved and then stay involved. Often, the only way to achieve this is to reimburse stakeholders for their loss of earnings and expenses.

Stakeholder dialogue is a specialist area of work and the skills necessary to ensure its success are not always present in MOD. Such skills are available in the private sector and can be bought in. A key requirement for stakeholder dialogue is to have the process designed in such a way as to be fair to all parties and to have a mechanism for resolving the inevitable disputes within the dialogue. These requirements are best met through the use of an Independent Facilitator.

IMPORTANCE OF INDEPENDENT FACILITATORS

By their very nature and name, Independent Facilitators are independent. They have no interest in the outcome of the dialogue or in the content of the discussions. Their role is to ensure that the dialogue is managed is such a way that it is more likely to achieve consensus.

Stakeholder dialogues have many means of failing, but the most common is that those stakeholders outside the usual company or government interests can perceive the dialogue as "unfair" or "rigged". The facilitator, by being an independent arbitrator and controlling the process (but not the content) of the discussions ensure that the process is fair and is seen to be fair. This allows stakeholders to participate fully in the process rather than withdraw. The withdrawal of stakeholders fundamentally compromises the dialogue process.

The other difficulty is in obtaining a sufficient range of stakeholders to make the process a legitimate public participation. Where NGOs and CBOs may be reluctant to participate in a dialogue if approached by the "defence" sector, they may be happy to attend when approached by an Independent Facilitator. Furthermore, Independent Facilitators have a large database regarding the correct people to approach within a NGO or CBO. They are better placed to get the right balance of participants for a stakeholder dialogue.

UK MOD EXPERIENCE

The UK MOD is developing its experience in using public participation methods for decision-making, particularly in the area of environmental management. It has begun using a stakeholder dialogue to resolve some environmental problems and participates as a stakeholder in national dialogues. Examples of these are given below.

ISOLUS - This stands for the **I**nterim **S**torage **O**f **L**aid **U**p **S**ubmarines. It is a MOD sponsored full stakeholder dialogue aimed at determining a suitable management method for redundant nuclear powered submarines that have had their nuclear fuel removed. It is independently facilitated by Lancaster University and full details of Phase I of the project can be found on the Lancaster University web site (www.lancs.ac.uk/users/csec/isolus/isolus.htm) MOD is in the process of seeking financial approval to begin Phase II. The ISOLUS Project has been recognised internationally as leading the field in public consultation at present.

Portreath - RAF Portreath in Cornwall was formerly used for the development of chemical weapons. This work ceased in the 1960s. As a result, the site is contaminated and it has been decided that it will be remediated. The remediation budget and timescales have been fixed, but MOD has begun a public participation process on the disposal of the waste and the traffic routes to be used. This is therefore a MOD bounded dialogue / Consultation.

SAFEGROUNDS - SAFEGROUNDS is a process examining the remediation of contaminated land on nuclear licensed and defence sites. After producing a technical document relating to the assessment of sites, the process went on to consider the management of contaminated lands. Five key principles were developed through extensive stakeholder dialogue and workshop processes. The stakeholder dialogue was commissioned through the Construction Industries Research and Information Association and was independently facilitated. All documents produced by the SAFEGROUNDS team are available for free download from www.ciria.org.uk through following the links to the SAFEGROUNDS project. MOD was a key stakeholder in the process, and the key principles have been widely accepted. The challenge is now for MOD to implement the principles - one of the principles required extensive stakeholder involvement.

MOD also participates in the BNFL national stakeholder dialogue and are members of the Business Futures Working Group of the dialogue. This again involves a wide variety of stakeholders and attendance is primarily to enable MOD to benchmark its processes with those in industry. The dialogue also serves as a useful source of information.

CONCLUSIONS

The UK MOD has developed considerable experience in stakeholder dialogue. This experience has helped stakeholders, particularly those who may otherwise be opposed to MOD on principle, to perceive MOD as an open organisation willing to discuss environmental matters and listen to stakeholder concerns. In this respect, the UK MOD is seen as being more effective that other Government Departments.

This situation only exists because MOD personnel attend a variety of stakeholder dialogues. The effect of this is to give the Department a "human face" and identifies a person to whom concerns may be addressed. This is not always the right person, but they can nevertheless act as a conduit into the organisation as a whole. This results in the breakdown of oppositional views, and increases the probability that the public at large will accept the MOD viewpoint.

However, it must be recognised that stakeholder dialogue is a slow and expensive process. Before embarking on a stakeholder dialogue, sufficient time and resources must be allowed for the process to be effective. It must also be recognised that bad stakeholder dialogues or public involvement processes are more damaging than no process at all. Great care must be taken in selecting the type of public participation (consultation, stakeholder dialogue etc.) required to avoid damaging dialogues.

Effective public involvement processes can be put in place provided the ground rules are clear - the expectations of the public must not be raised beyond the ability of the process to deliver the output. This is at its most important in bounded dialogue conditions, where the boundaries and constraints must be clearly identified.

Where it is intended to use bounded dialogue or full stakeholder dialogue, it is critical to use an independent facilitator. They have the skills to ensure that a dialogue does not collapse due to their position of being independent from the outcome. Stakeholders are also unlikely to consider a dialogue to be fair where it is facilitated by an organisation having a stake in the outcome.

Taking all these factors into account, the UK MOD considers public participation is a key method of operation in the current political and societal climate. Public involvement should take place at an appropriate level and in an appropriate manner. The decisions on what is appropriate must be retained within the MOD.

Environmental Policy in the Slovak Armed Forces, Environmental Management System as a Challenge for the Future

Lubomir Kušnir,
Infrastructure and Environment Protection Division, Ministry of Defence of the Slovak Republic, Bratislava, Slovakia

The State environmental policy in the Slovak Republic was created in 1993 as an integral part of a new government policy. This policy has followed European legislation in this area and identified short, medium and long term objectives. The military sector did not have any exceptions from requirements of State Environmental Policy. To meet all requirements of State Environmental Policy, the Conception of Environmental Protection within the Armed Forces was developed and approved by the Minister of Defence in 1996. The content of this Conception is based on principles of State Environmental Policy and includes short, medium and long term objectives directed towards specific military conditions. Along with state environmental legislation this Conception is one of the basic tools in this area within the Armed Forces. The approaches and experiences of NATO countries have been incorporated into the defence environmental policy.

A great deal has been accomplished in Defence environmental policy as part of the preparation of the Armed Forces for NATO membership. The Government of the Slovak Republic approved a national programme "Preparation for NATO Membership" in October 1999. Major government authorities were directed to support this effort within its area of responsibility. The Ministry of Environment established a working group from representatives of the Ministry of Environment, Ministry of Defence, Ministry of Health, and the Ministry of Agriculture. This working group identified two major projects needed to promote environmental protection within the Armed Forces. The implementation of Environmental Management Systems (according to norm ISO 14001) to the Slovak armed forces is the first project. The second project is Development of a Pollution Sources Database for Military Installations. Initial studies on both projects were approved by the Minister of Defence in 2001. The realization phase of these projects has also begun in this year.

The Environmental Management System (EMS) had started with all proceedings according to norm ISO 14001. Identification and

K. Mahutova et al. (eds.), Defense and the Environment: Effective Scientific Communication, 71–74.

prioritisation of environmental aspects of 12 selected installations represented major military activities (air bases, training areas, fuel storage, ammunition depots, refueling stations, etc.) was done in 2001. The working group has decided that the most representative case of EMS according the norm ISO 14001 will be completely to implement an EMS at one selected military installation. In 2002 Air Base Kuchyňa was selected for this purpose. The procedure included EMS education in all levels of air base management, developing requested forms of documentation according to norm ISO 14001, developing and statement of an Air Base Environmental Policy, negotiations with the local environmental state authority, and carrying out an internal audit. All steps have been done to prepare the military airbase for an external audit on EMS. The working group has decided that the largest military fuel storage facility in Slovakia – Zemianske Kostol'any is the most appropriate installation where an EMS will be implemented in 2003.

The database of sources of contamination in military installations is an integral part of government objective to develop databases of pollution sources within the Slovak Republic. The military sector is the first government authority which will fulfill this objective. The development of the database is an inevitable condition for the objective decision making process of clean up activities. The system of database elaboration is based on the Danish software system GeoEnviron which has been used in the Danish army. The system was modified to be Slovak specific and military sector specific. The database does not only include data. There is also potential for risk evaluation of the site and a scoring system to prioritize measures. The risk analysis is a very significant part of the database. This comprehensive information will allow an objective decision making process. The intention is to develop databases for all military installations within three years. The database of military installations from the east part of Slovak republic was developed in 2002. The central part of Slovakia was in preparation in 2002. Development of the database's final part will be for military installations in the west part of Slovakia in 2003. After fulfilling this objective, the database of all military installations will be in place. A completed database allows the allocation of financial resources for necessary clean up activities to localities where the most significant risk to the surrounding environment exists.

The new European Union legislation incorporated within the legal system of the Slovak Republic also essentially has changed Slovak environmental legislation. The new acts on Nature Protection, Waste Management, Water Management, Protection of Air, etc., approved by Parliament mainly in 2001 and 2002 very significantly changed conditions and duties for the Armed Forces in environmental protection. To execute these obligations the Ministry of Defence issued a very significant document in 2002. The Minister of Defence issued a new order addressing environmental protection within Armed Forces. The Director of Modernisation and Infrastructure Department of the Ministry of Defence issued Guidelines for Environmental Commissions in all military installations and Guidelines for Environmental Measures in the military installations.

The most significant document issued in 2002 is „Armed Forces of the Slovak Republic - Model 2010“ prepared for the reconstruction of the Armed Forces in order to join NATO. This document very essentially identifies all aspects and missions of Slovak Armed Forces till 2010. Environmental protection is one chapter of this document which will be valid until 2010. This chapter identifies environmental protection objectives, measures and responsibilities for all persons within the Armed Forces.

The defence environmental policy will reflect both national and NATO requirements on environmental protection. International cooperation within the framework of NATO committees, NATO working groups and other bodies will increase environmental awareness of the defence sector.

Based on last year's experience in implementation of Environmental Management Systems within the Armed Forces is the most appropriate environmental protection challenge. If the defence sector has in place Environmental Management Systems it is a message to the public that the Armed Forces take care of the environment. This includes top management committed to legal compliance, and continual improvement of environmental performance, the identification of objectives and targets, the creation of an environmental programme, and allocation of financial and human resources.

By implementation of Environmental Management Systems the defence sector will promote public awareness of their achievements and educate the public to their needs in the support of national security.

SUMMARY

The Environmental Policy in the Slovak Armed Forces is based on principles of the State Environmental Policy. The main objectives of defence environmental policy are protection of environment in the military units and facilities and during military training and operation. Decreasing measure of environmental impacts in most relevant indicators which has been indicated in results monitoring of air, water, soil and waste management is one of the most significant issues. Currently created conception, legislation, legal, economic and organizational conditions and systematic securing tasks strategy of defence environmental policy is base assumption for improvement of the environment in the areas of defence administration. Ministry of Defence closely cooperates with Ministry of Environment to fulfill objectives of the state environmental policy.

The most significant result of cooperation in framework NATO in environmental area is starting process of implementation of Environmental Management Systems to the Slovak Armed Forces. Ministry of Defence and Ministry of Environment had prepared conditions for this process reflecting both a current national focus and NATO activities. EMS as a systematic approach to the protection of environment is one of the most significant tools for this process in special armed forces conditions.

Environmental Management Systems (EMS) in the Military Sector
Promoting Broader Implementation

Michael Dawson
Environmental Management Systems and Sustainable Development Strategies, Department of National Defence, Ottawa, Canada

INTRODUCTION

The EMS is the framework of procedures, processes and practices designed to help an organization manage its environmental agenda and document, evaluate, and communicate its environmental performance.

In broad terms, the ISO standard prescribes a five-phase cycle of Policy, Planning, Implementation and Operation, Checking and Corrective Action, and Management Review that is designed to result in the continuous improvement of the organisation's environmental performance.

It is important to stress that the environmental programme is established by the organization itself. This means that the organization, in developing its EMS, identifies the significant aspects of those products, activities and services for which it is responsible and over which it exercises control and develops plans to address them all within the confines of the organisation's environmental policy.

THE CCMS PILOT STUDY ON EMS IN THE MILITARY SECTOR

In March 2000, the NATO Committee on the Challenges of Modern Society (CCMS), Pilot Study on EMS in the military sector concluded in its Final Report (1) that, "it is possible and even desirable to implement environmental management systems in military organizations." Further, the pilot study group recommended implementing a standard for EMS "that is recognized "all over the world". (The only such standard currently available is the International Organisation for Standardisation (ISO) 14001 standard.)

The Pilot Study concluded that, "EMS at the very least, provide safeguards for top management to ensure that environmental legislation is complied with".

K. Mahutova et al. (eds.), Defense and the Environment: Effective Scientific Communication, 75–83.

The theory is both eminently sensible and – on the face of it - relatively simple to implement. The benefits are obvious both internally in terms of reduced liability and cost savings and externally in terms of bolstering a positive public perception of the military as sound environmental citizens and as an integral element of the larger society.

For an international, multi-operational entity such as NATO, the positive impact on interoperability is self-evident.

Given these advantages, why is there such a degree of organizational resistance?

The challenges are considerable. Implementing an EMS requires a clear policy from the senior management cadre that must constantly choose among competing priorities to distribute scarce resources. Middle management – already balancing the challenges of dealing with shrinking budgets and expanding task lists ("do more with less") must "champion" the EMS, translating the policy requirement into a pragmatic management framework that is understood and supported by every level of the affected organization. In turn, the operational level must incorporate environmental considerations into day-to-day decision-making by developing and implementing yet more and more rigorous procedural instruments (Routine Orders, Standing Operating Procedures etc.) and checks.

AN EXAMPLE OF EMS IMPLEMENTATION: THE CANADIAN EXPERIENCE

The endorsement of EMS frameworks for managing environmental issues continues to grow across the public and private sectors, encouraged by the pressure of public opinion and by the scrutiny of bodies like the Organisation for Economic Cooperation and Development and the European Union.

Nevertheless, as late as 1996, a risk management practice survey conducted in Canada found that crisis management was the most popular approach to addressing environmental issues. This is in spite of the fact that major environmental legislation such as the Canadian Environmental Assessment Act (2) and the Canadian Environmental Protection Act (3) was in force and binding on federal entities including National Defence and the Canadian Forces. Furthermore, the revised Auditor General Act (4) required departments and agencies including the Department of National Defence (DND), and the Canadian Forces (CF), to table in Parliament a strategy for achieving sustainable development within the organization and for updating that strategy at least once every three years.

To ensure implementation of these strategies, the federal Cabinet directed that federal departments and agencies implement EMS consistent with the principles of the ISO standard.

The first Sustainable Development Strategy (SDS) was tabled in Parliament in 1997 (5). To ensure that the targets contained in SDS '97 were implemented, DND/CF simultaneously undertook implementation of an EMS that conformed to the principles of the ISO 14001 standard across the organization. The second SDS (6), tabled in 2000, contained a target to implement EMS at every level of the organization from Headquarters to Bases and Wings by 31 March 2004.

DND and the CF began turning talk into action simultaneous with the convening of the CCMS Pilot Study. We have remained active and committed members to this day.

THE CCMS EMS WORKSHOPS

Late in the Pilot Study process, participants became aware that resistance to EMS was a serious enough threat to its implementation that follow-up work was needed.

Thun: 2001

In January 2001, in response to a recommendation contained in the Pilot Study Final Report, a workshop on EMS was convened in Thun, Switzerland. The goal of the workshop was to bridge the gap between the theory of EMS and the practice (i.e. implementation, especially for those organisations that were only beginning to consider the implications of adopting an EMS approach to environmental management). This workshop resulted in the promulgation of complementary guidelines (7) for EMS implementation. Among the notable conclusions of the workshop:

- (i). The environmental policy must balance environmental protection with the military mandate;
- (ii). The speed of implementation varies dependent on societal pressures, political leadership, culture, organizational structure, the legislative framework, etc.;
- (iii). In most countries, at least some elements of an EMS already exist;
- (iv). Implementation of EMS with other management systems is both efficient and effective; and,
- (v). Implementing EMS within the military is a national effort; however, its context is international.

Conclusions (ii) and (v) are the particular focus of the recommendations for the future direction of CCMS EMS workshops.

Vyskov: 2002

In September/October 2002, CCMS, the (Canada) Department of National Defence and the (Czech Republic) Military University of the Ground Forces, in Vyskov co-sponsored the second EMS workshop, co-hosted by Canada and the Czech Republic.
In the course of that workshop, a number of issues that had come to light during the Thun workshop, particularly Thun conclusions (ii) and (v), were addressed in greater depth. As a result of the Vyskov workshop, a series of recommendations was put forward in a report (8) to the CCMS EAPC-format plenary meeting of March 21 2003 at NATO Headquarters in Brussels.

The Vyskov Diagnosis

As noted previously, the pace of EMS implementation is varied both in terms of the individual implementer's capacity to address environmental issues and in terms of experience in the implementation of management systems.

Participants from Albania, Austria, Belgium, Canada, Czech Republic, Denmark, Georgia, Germany, Italy, Kyrgyz Republic, Latvia, Lithuania, the Netherlands, Norway, Romania, Spain, Sweden, Switzerland, Ukraine, the United Kingdom, and the United States of America attended the Vyskov workshop. The key findings of interest are highlighted in bold.(The full text of the technical report is attached hereto as an appendix.)

Anecdotal evidence suggests that some of the participants in the original pilot study did not attend the Vyskov workshop because they had abandoned plans to fully implement an EMS; however, without further research it is impossible to confirm this possibility or to identify an office of primary interest (OPI) with whom to follow up.

Overall, there was clear consensus that:

- measurable progress in implementing EMS within the military sector had been made since Thun and generic solutions to some of the more common problems are feasible;
- there are common issues facing EMS practitioners across the spectrum of experience and knowledge whether in defining significant environmental aspects (e.g. waste reduction, water and energy conservation or contaminated sites), addressing conflicting policy stances (e.g. inability to pursue formal ISO registration due to variances in the ISO audit requirements and established security policy), or developing meaningful performance measures and reporting processes;

- This generic nature of problems and solutions is a key building block toward interoperational EMS practice in the military sector;
- The need for a future forum to address performance measurement and reporting was almost unanimously accepted; and,
- The **gap in knowledge among participating organizations is widening rather than narrowing**. (Participants were able to identify three broad tiers of EMS capacity: beginners; intermediate; and, advanced.)

SYSTEMIC BARRIERS TO EMS IMPLEMENTATION

An Established Legislative/Regulatory Framework

A distinct advantage to the implementation of EMS within any organisation is the existence of an externally imposed legislative and regulatory framework. The newly emerging (i.e. Tier 3) democracies face a number of challenges. They must put in place durable political institutions that in turn must establish a legislative and regulatory framework that is built for the most part "from square one." In other words, while NATO's original member nations have built their environmental agendas on an evolutionary model, the newer members are working a revolutionary model where not only are the proposed environmental laws new but the governance structures, regulatory agencies and delivery mechanisms within the affected organisations are in development simultaneously. Add to this the critical financial constraints, the significant legacy issues (e.g. contaminated sites clean-up) and the competition among critical interests for recognition and programme funding and the task of the EMS practitioner becomes truly daunting!

A significant difficulty in implementing EMS within the military organisations of the emerging democracies is that the external legislative and regulatory framework that emanates from the legislative body is often immature and/or incomplete. This fact makes it difficult to implement ISO-based EMS because – in the absence of clear legislated requirements for addressing environmental issues - the top managers take a short-term view of those issues primarily as a financial burden that drains already scarce resources rather than as a strategic cost benefit.

A Pragmatic Interpretation of the ISO Standard

Experience with EMS implementation has shown that a failure to grasp the pragmatic implications of the framework itself cause the implementing organisation either to duplicate effort or to abandon the process of implementation as hopelessly complex.

The standard was developed primarily with private sector enterprises in mind. It is weighted toward site-specific operations. This can cause difficulties.

For example, the ISO standard refers to "top management". Within the government of Canada, officials in most departments mistakenly assumed that this meant the Minister/Deputy Minister in charge. EMS implementation subsequently bogged down because they could not identify existing processes that involved the Minister/Deputy Minister directly or personally.

In fact, "top management" must be interpreted as the position in the organisation that can make a decision critical to any element of the management system. For example, "top management" in financial matters within National Defence is the Programme Review Board because it has final say in all expenditures. While other departments in the federal government were trying to create processes to involve the Minister as "top manager", DND/CF recognized the multiple nature of top management and was able to take advantage of processes and procedures already in place when analyzing the gaps between the existing and the ISO-prescribed management frameworks. The vital message here is that military organisations, by the very universality of their nature, already have in place on average 90% of the elements of an ISO environmental management system.

REMOVING THE BARRIERS

EMS is fully implemented when environmental considerations are fully integrated into the decision-making process at every level of the organisation. Key to successful implementation is the endorsement of EMS by the critical players (the "top managers") at each level and within each function.

Interestingly, the recommendations from Vyskov were developed independent of but in close alignment with a similar mentoring partnership programme (the Individual Partnership Activity Programme (IPAP)) being proposed centrally at CCMS.

Both IPAP and the Mentoring Partnerships proposal developed at Vyskov propose cost-effective and pragmatic opportunities for pursuing effective knowledge transfer. The logistics of the approach are fairly straightforward and readily accepted. In the case of the Vyskov workshop these include:

- consultations to identify needs and to identify existing "centers of excellence";
- establishment of a web-site featuring an electronic library, a list of subject specific experts willing to share their expertise, and a "chat room" where emerging issues can be explored by interested practitioners;
- a proposal for (a) further workshop(s) possibly featuring "hands on" demonstrations of the systems in place; and'

- a bilateral mentoring program whereby first and second tier MOs can work together with third tier MOs to identify specific needs, and develop plans for the transfer of knowledge and technology needed to answer those needs on a one on one basis.

To ensure that all MOs have an equal opportunity to avail themselves of a functioning EMS that puts them on an equal footing with their NATO and EAPC counterparts, mentoring one on one is the optimum solution. The only significant missing element is the aforementioned "legitimacy" factor wherein senior management needs external validation in the form of legislation or regulation to justify the expenditure of scarce resources on environmental activities. In the absence of robust legislative regimes, this matter can be addressed through a concept that we will call "Environmental Health Indicators".

ENVIRONMENTAL HEALTH INDICATORS

It is a given that certain factors are vital to the sustainability of a society. Top priority among these factors are clean water, pollution prevention, energy use and conservation, clean air, and – for the military – useable (i.e. sustainable) training areas.

The military sector is an integral element of the broader society it serves. It shares water and air with the rest of society, not to mention the world. Given the global preoccupation with environmental degradation and its side effects on economies and societies, there are numerous standards against which to measure a number of environmental indicators relating to water, air, and pollution. The Organisation for Economic Cooperation and Development, the United Nations, the International Fund, and the European Union to name a few, use standards of varying degrees of precision to measure these things. The point is that the standards are available. Therefore, even for military organisations in countries where the legislative and regulatory framework is not sufficiently developed to allow the military organisation either to adopt a national standard or to be subject to national legislation, a standard suitable to the capabilities of the military organisation interested in pursuing EMS can be either adopted or adapted with a full measure of credibility.

While the standards and performance indicators for military training areas are less exact, work on this issue and the associated benchmarks has been done and is available to be shared.

In simple terms, environmental health indicators for air soil and water can be developed by any military organisation to reflect its environmental priorities pragmatically and to allow that orgainsation to integrate environmental management into its existing management framework without waiting for the broader society's political institutions to build the legislative regulatory and policy framework that at first glance appears necessary to support the EMS.

THE WAY AHEAD

There is little value in discussing the financial implications of EMS implementation in the military sector for a number of reasons, first and foremost because an accurate cost estimate is simply not feasible. For one thing, ascribing the environmental cost element to an activity like replacing outdated plumbing fixtures is so arbitrary as to be pointless. It is fair to argue that EMS costs only staff time. The shortcomings that the analysis of the standard against the existing management system brings to light were identified rather than caused by the move toward EMS. Finally, both documented and anecdotal evidence suggest that the cost implications of EMS implementation are not all negative. Some items are revenue saving, revenue neutral or even revenue producing (e.g. fuel storage tank rationalization results in the sale of excess capacity).

Nonetheless, military commanders in today's climate of downsizing, expenditure reduction, encroachment on training areas and the often negative public perception of the military cannot be expected to embrace an innovation that - on the surface at least – appears more sentimental than rational.

For the reasons discussed previously in this presentation, the commanders of military organisations within the emerging democracies face far greater challenges in embracing an EMS.

There is no doubt, however, that the one-on-one mentoring partners approach recommended by the EMS working group specifically for EMS implementation and supported by the broader CCMS agenda for Individual Partnership Activities Programmes is the most cost-effective means of bringing a common approach to bear on environmental/sustainable development issues among the NATO and EAPC countries' military organisations.

The next step is of course to convince potential partners to adopt this way of doing business. As an example of a "work in progress", Canada, Sweden and Lithuania are collaborating on EMS implementation within the Lithuanian Ministry of Defence.

A discussion among those here present would identify numerous such collaborations, some more formal with detailed Memoranda of Understanding, others more informal and with perhaps less defined objectives. The fact remains that the opportunities are plentiful for developing targeted, subject-specific, time-bound and successful partnerships that take full advantage of the current political atmosphere of cooperation in place of confrontation.

THE FINAL PARADOX

In closing, it is worth noting that those of who advocate less structured and more "one-on-one", collaborative approaches to addressing common concerns outside the fora of NATO and the CCMS are nevertheless utterly reliant on those fora to deliver the message. Sessions such as this ARW, the EMS pilot studies and workshops, the annual EUCOM conference and CCMS plenary meetings allow innovative approaches like mentoring partnerships to be proposed, debated, and developed – mostly on small budgets and supported by the huge energy reserves and enthusiasm of the proponents.

REFERENCES:

1. *Environmental Management Systems in the Military Sector, Final Report of the Pilot Study Group No 240 (March 2000)*
2. *Consolidated Statutes of Canada Government of Canada website (www.gc.ca)Justice: Canadian Environmental Assessment Act*
3. *Consolidated Statutes of Canada Government of Canada website (www.gc.ca)Justice Canadian Environmental Protection Act*
4. *Consolidated Statutes of Canada Government of Canada website (www.gc.ca):Justice Auditor General Act*
5. *SDS '97 Government of Canada website: (www.gc.ca):National Defence:ADM(IE)*
6. *SDS 2000 Government of Canada website(www.gc.ca):National Defence: ADM(IE)*
7. *Environmental Management Systems in the Military Sector; complementary Guidelines for the NATO/CCMS Final Report 240, (240B) January 2001*
8. *Implementation of Environmental Management Systems (EMS) in the Military Sector: NATO Report 258*

A Corporate EMS for the UK Ministry of Defence

Roger T. Moore
Ministry of Defence, London, United Kingdom

INTRODUCTION

The central policy drive within the MOD is to put in place and promote policies that are cost effective to implement, protect defence assets, and protect the environment. Closely linked to the environmental agenda is a UK government strategy to put sustainable development "at the heart" of government [1]. To oversee this work and monitor progress, committees of ministers and officials have been set up. The UK environment ministry, DEFRA, developed a set of sustainable development indicators [2]. Parliamentary scrutiny of progress to meet these targets is being carried out by the Environmental Audit Committee. To ensure independent review of the policies and their implementation, the government has set up a Sustainability Commission [3]. Central to these policies is the need to provide leadership, implement environmental management systems in all government departments, "green" the operations of government. [http://www.sustainable-development.gov.uk]

TARGETS

To provide a basis for comparison and reporting, a number of targets were developed [4]. These included:

- a 1% reduction of CO_2 per year,
- 10% of all energy purchased would be from renewable sources
- benchmarking actual energy use against best practice and set targets
- recycling 40% of office waste
- reducing water usage to under 8 cubic metres per year
- reducing fuel consumption by 10% by 2006
- buying recycled paper and certified timber

K. Mahutova et al. (eds.), Defense and the Environment: Effective Scientific Communication, 85–89.

ENVIRONMENTAL MANAGEMENT SYSTEMS IN GOVERNMENT

There are a number of standards available which can be used to develop effective environmental management systems, these include: ISO 14001 [5] and EMAS [6]. They all include the basic principles of any management system with the emphasis on leadership, initial review and continuous improvement.

The Government has decided that all central government ministries dealing with, for example, employment, trade and industry, food, police, prisons, must implement environmental management systems based on ISO 14001 or EMAS 2 or similar [7]. The timetable set down for their introduction is as follows:

- all main offices or 40% of the estate by 31 March 2004
- all offices or 80% of the estate by 31 March 2006

The targets are set to allow for the differing make-ups of Departments. It is also important to note that there is no mention of certification. Although some areas may feel there are benefits in seeking third party certification, it is not a mandatory requirement.

MOD EMS DEVELOPMENT

To implement the environmental management systems in the UK MOD an EMS team was formed and an implementation plan developed. The plan covered the UK and MOD bases overseas. It was decided to base the system on ISO 14001. To enable areas within the MOD to develop a coherent approach, guidance and a manual were produced and approved at a senior Navy, Army and Air Force. So that the systems could be checked for efficient operation they were integrated with the SHEF (safety, health, environment and fire) audit process which itself is based on the principles of ISO 14001. Appropriate training and software were employed to underpin the work.

To illustrate the scale of the undertaking I will describe the Army's approach. The Army decided that they did not have sufficient internal expertise to deliver to the required timetable so they employed consultants to run on-site courses targeted at selected personnel. The Army developed a 5 year plan. The first year focussed on trials of the system; in the next three years it was proposed to cover 200 sites each year and in the final year sites and buildings used by reservists and cadets. This included sites overseas. It can be seen that this was not a trivial task either in its planning or execution.

Other parts of the MOD are at various stages of implementation. The Navy are building on a pre existing compliance system. The Marines are to adopt a plan which involves external certification for energy management alone. For the RAF a programme of implementation base by base is being adopted. This is now 75% complete. Headquarters buildings in London are being dealt with by a commercial contractor as part of a 'privatization' exercise. A more difficult challenge is faced by the Defence Procurement Agency in trying to adapt the EMS to fit their business processes which are essentially not site based. Work is in hand and trials are being run with a few projects. The Defence Science and Technology Laboratory DSTL has built an integrated management system to address all corporate risks. Defence Agencies that see a commercial advantage in marketing their services have sought third party certification - eight in total.

ASSESSMENT OF SIGNIFICANT ENVIRONMENTAL ISSUES

In running the management system there is a need to identify issues which will be measured and examined, for the MOD these included:

- Air emissions
- Discharges to water
- Waste
- Land contamination
- Energy and fuel use
- Water use
- Use of other resources
- Nuisance
- Bio diversity
- Public interest
- Legacy
- Planned changes
- Travel
- Purchasing

The underlying philosophy behind the MOD EMS and its implementation was one of simplicity. Keep It **SIMPLE** was the catch phrase. On this basis **S**ignificant environmental issues were to be addressed first. The system was to be **I**ntegrated into everyday business. It was seen as key to ensure that it did not introduce an additional burden or created confusion and potential conflicts. It was essential that the system was **M**onitored to check on performance and to identify savings made and potential future savings. For the system to yield the benefits required and for it to survive it needed to be accepted and be **P**eople driven. Concentration on behaviour and education was seen of central importance. A system of this type cannot be imposed with any lasting success; everyone has a part to play. Gaining continuous improvement relies on the willingness of all to participate and think widely.

LESSONS LEARNED

Environmental management systems are as much about good management as they are about protecting the environment.

Experience has shown that people are usually keen to support the introduction and running of the system. For most things the 80/20 rules applies; that is that limited effort will yield significant gains when targeted on the correct things but that improvements become more and more difficult once the easy gains have been achieved. So it has been important to emphasise the need to keep matters practical and focused on those areas that will, in the first instance, yield maximum benefit. It was mentioned earlier that leadership and organizational commitment were key to effective implementation; this has been found in practice. The ability to accrue knowledge and information from all sources and use it to deliver good performance is important. Using what already exists, learning from others and evolving through team work will deliver real benefits. And it must be always remembered that EMS is just a tool and not an end in its own right.

SUSTAINING UK DEFENCE INTO THE FUTURE

There is no doubt that the pressures on defence to take notice of the environment in which it operates are increasing. There is a feeling that we will need to be proactive or face future constraints. These could include limitations on training caused by noise restrictions, controls on the access to sites and difficulties associated with meeting the needs of conservation. Restoring land to a level deemed acceptable by outside agencies be they official Regulators or NGO pressure groups can be very costly. Also, if badly managed, there is the potential for substantial compensation claims.

Financial pressures in the future will increasingly create the need for efficient use of resources – and the defence estate is a critical and valuable resource. Another challenge will be the need to ensure that a capable and effective workforce is maintained in order to be able to manage the systems and deliver the results required. Effective defence requires public support and the avoidance of political embarrassment. A well managed sustainability strategy will underpin that aim.

Although rarely considered explicitly, Defence provides a key part of sustainable development. Without a stable and safe society, the other elements of sustainability would come to nought. It could be considered that the various elements and issues are bounded by four concepts: sustainable development, governance, risk, and corporate social responsibility. The milieu includes economic, social and environmental matters.

So what does all this mean to the MOD? The EMS and the linked performance indicators set by Ministers for, for example, water, waste and biodiversity will form a key element in the strategy. The drive for continuous improvement, and not just legal compliance, is supported by the reporting systems and improved data collection. For social issues, families and communities will get increased support. Health, health and safety, lifestyle screening, fitness facilities are all parts of evolving plans to maintain and improve corporate health. Coupled to these programmes are human resource innovations to provide diversity, a better work life balance, support to parents with young children eg child care facilities and a variety of flexible working patterns to match the requirements staff. Many of the initiatives associated with economic matters are closely linked to procurement programmes, government regional and local planning and its direct and indirect impact on employment and the local economies and decisions about site closures

The whole purpose of the armed forces and is to provide peace and security. The central role is as a force for good.

CONCLUSIONS

It can be seen that the environmental management system has been described in the context of a broader government sustainable development policy. The EMS provides a means of effective delivery of a major part of the strategy. For the MOD, key elements include: environmental impact appraisal of significant policies and programmes; introduction of work streams to develop guidance and procedures and the definition of short term targets and long term goals. Leadership, being best in class and benchmarking are all matters of importance.

There has been recognition that the creation of opportunities to engage with a range of stakeholders may have great benefits. Being open and transparent coupled with better reporting, improves general awareness and lessens public concern.

REFERENCES

1. 1997 Labour Manifesto Commitment
2. A better quality of life – a strategy for sustainable development for the UK (May 1999)
3. The Commission held its first meeting in London on 23 November 2000. It reports to the Prime Minister and the devolved Administrations.
4. The first parts of the Framework for Sustainable Development on the Government Estate were published on 23 July 2002. This is the main vehicle for improving the performance of the Government Estate and will eventually cover all key environmental and social impacts of the running of departments. The Framework also contains guidance to support Departments in achieving targets.
5. See International Organisation for Standardization website for further details, http://www.iso.ch,
6. See http://europa.eu.int/comm/environment/emas/about/summary_en.htm
7. Framework for Sustainable Development on the Government Estate Target A3.

Reccomendations based on the NATO-CCMS Pilot Study on Methodology, Focalization, Evaluation and Scope of Environmental Impact Assessment.

Katarina Pavlickova
Dpt. of Landscape Ecology, Faculty of Natural Science, Comenius University in Bratislava, Slovakia

INTRODUCTION

Environmental Impact Assessment (EIA) is a key part of the development process for all large projects in the many countries around the world. An EIA is required for all projects located close to military areas.

The requirements for EIA within EU are set out in Council Directive 97/11/EC of 3 March 1997 amending Directive 85/337/EEC on the assessment of the effects of certain public and private projects on the environment. The two Directives have to be read together. These Directives are considered as a basis for national legislative requirements within the EU and applicant countries.

The main principles of EIA under the amended Directive are that it should take place for private and public projects likely to have significant effects on the environment, and that development consent will be dependent on an adequate EIA having been undertaken. Separate categories are defined for those projects which will always require environmental assessment and those for which environmental assessment will only be required at the discretion of the Member State. The Directive specifies minimum requirements for the information which should be contained in the EIA report and advice is given on how to make the decision as to whether an EIA should take place, a decision process called "screening" [2]. After screening, the scoping stage and other steps of EIA procedure establish the issues that are to be covered by the EIA according to national legislation.

INITIATION AND DURATION OF THE PILOT STUDY

The NATO-CCMS pilot study on "methodology, focalization, evaluation and scope of environmental impact assessment" was launched in November 1991. At the beginning NATO member states officially

K. Mahutova et al. (eds.), Defense and the Environment: Effective Scientific Communication, 91–100.

participating in the pilot study were Luxembourg, Greece, Italy, Portugal, Spain, Turkey and the United States. During the time, other countries were involved: NATO members and the official participating countries, environmental impact assessment (EIA) experts from France, the United Kingdom, Canada, Norway, Iceland, Poland, Lithuania, Bulgaria, Slovenia, Finland, Hungary, Czech Republic, Slovakia, Russia, Rumania, Estonia, Austria and Togo. The director of the pilot study was Prof.Dr.R.F. Verheyen of the University of Antwerp. The University of Antwerp played the role of leading management organization.

The pilot study was initiated by the Belgian Office for Scientific, Technical and Cultural Affairs (DWTC) for a period of three years. During that period seven workshops were organized. Due to the success of the pilot study, the participants' strong demand for more information exchange on several aspects of EIA, and the international request for study of more EIA aspects, the pilot study was extended twice. It was closed at the beginning of 2002. Sixteen workshops have been organized throughout the project; the final workshop was organized in April 2002 [3].

The pilot study was based upon regularly organized workshops dealing with specific themes. Workshop participants were requested to gather information on the selected theme and to present it at the workshop. The information of different countries were compared and discussed. The contributions were complied into reports on several aspects of EIA [3].

OBJECTIVES OF THE PILOT STUDY

The pilot study aimed to discuss and compare different aspects of EIA, and to improve the EIA process by giving recommendations to the concerned authorities. Many differences between the participating countries occurred related to procedures, scope, methodologies, depth of supplied information, and evaluation criteria used in environmental impact assessment. Workshop participants not only were from different countries, but also had different fields of experience and represented administrations, universities, agencies, NGO's, EIA centers, as well as environmental offices. The pilot study brought together these different views on and experiences with environmental assessment (3).

Final reccomendations are concerned with the themes of quality control; effectiveness of EIA; public participation; land use planning and Strategic Environmental Assessment (SEA); water management and EIA; transportation and EIA

QUALITY CONTROL IN EIA [9]

Quality control in Environmental Assessment (EA) needs to give attention to at least two facets: the evaluation of the process, and the evaluation of the final documents, such as the Environmental Impact Statement (EIS).

Quality can be described as "characteristic" of goods in relation to the use made of it. Taking into account this definition means that the quality of an EIS depends on the users of the EIS. The EIS needs to meet all quality criteria from the different users.

The developer is interested in a short EA process, reaching a project authorization and carrying out mitigation measures or project changes at the least cost. The consultants are interested in 'easy' and repeatable (reproducible) methods for impact prediction and correct baseline data. The authority needs to check the procedural compliance, give accreditation to experts, and review the EIS.

The pilot study results lead to the following recommendations on quality control in effectiveness in EIA:

1. Give more regard to the management skill.
2. Public tenders with a strict description of qualifications, appropriate to the type of project.
3. Provide information to developers on aims and prerequisites of EIA before the procedure.
4. Develop lists of consultants using the criteria of best practice, good practice and not-suitable practice. There needs to be a clear process of accreditation and free competition between different experts.
5. More time is needed for scoping.
6. Ensure that manuals have widespread distribution and are kept up-to-date.
7. Provide training related to practical skills
8. Engage independent experts as support mechanisms in all stages of the review process.
9. Continuous guidance during the complete "EIS process" by a (technical) steering committee of the consultant's bureau.
10. Post evaluation of the EA process is necessary.
11. The influence of the public through suggesting new alternatives, using specialized (residents) knowledge on local areas or specialized items.

PUBLIC PARTICIPATION [5]

Public participation in the EIA process is an element which guarantees environmental safety. Public inquiry can be defined as the informative phase for citizens living in the involved area and an opportunity to express their opinions. It is also useful when technical-scientific aspects can joint the participation of the community involved. Effective public participation is defined as follows:

- Identifying public concerns and values,
- Gathering economic, environmental, and social information from the public,
- Keeping the public informed about potential actions or alternatives, together with potential consequences,
- Developing and maintaining credibility, and,
- Ultimately, improvement of the decision-making capabilities of the agency (4).
- Rights to public participation in environmental decision-making are based on one of the following regulations:
    - various constitutional rights,
    - specific environmental or nature protection laws, including EIA, licensing, permitting, land-use planning, etc.,
    - Administrative laws or codes.

During the course of pilot study we focused on the second group with these recommendations for the increasing of public participation:

1. Expand the transparency of the EIA and SEA process and the decision-making process.
2. The participation rights should be as broad as possible, and should include the right to be informed, the right to be involved at an early stage, and the right to be heard.
3. Provide guidelines on the role of the public in the EIA and SEA process.
4. Provide targeted training on public participation to the different "actors" involved in the EIA, SEA and decision-making process.
5. Ensure more interaction of the public participation process; cooperation at all EIA and decision making stages.
6. Introduce strategic EIA and public consultation at earlier stages than the project level.
7. In the case of transboundary situations, give attention to differences in cultural habits and temper, when organizing public participation in the neighbouring countries.
8. Provide more support to enable NGOs to meet the technical and quality standards required to represent public opinion in an appropriate way.
9. Countries should further develop practical procedures for early and effective involvement of the public.
10. Public participation in licensing and permitting of specific activities is quite limited in most countries, with participation

often limited to the "affected" public. There are inadequate mechanisms to ensure that comments are seriously taken into account.

11. Adopt the definition of "the public" and "public concerned" as establish in the Aarhus Convention.

STRATEGIC ENVIRONMENTAL ASSESSMENT (SEA) [1]

This part of the pilot study dealt with strategic environmental assessment of sector strategies. Relating to these, the EU SEA Directive (Directive 2001/42/EC of the EP and the Council of 27 June 2001 on the assessment of the effects of certain plans and programmes on the environment) lists in Article 3(2), plans and programs (PP's), "which are prepared for agriculture, forestry, fisheries, energy, industry, transport, waste management, water management, telecommunications, tourism (...) and which set the framework for future development consent of projects listed in the Annexes I and II to Directive 85/337/EEC (...)."

The SEA topic is an emerging and important one, because all EU Member States have to authorize the laws, regulations and administrative provisions necessary to comply with the SEA Directive before 21 July 2004. In the light of the enlargement of the EU to 27 countries, all these countries are requested to implement the Directive's provisions in their national legal systems. Consequently, the Directive itself has world-wide impacts and affects. on the environmental legislation of developed countries outside the EU (e.g. Japan), of transition countries and so-called developing countries, and on the environmental policies of international organizations (e.g. European Bank for Reconstruction and Development (EBRD), United Nations/Economic Commission for Europe (UNECE), because it is acting as a state-of-the-art model for SEA.

It is necessary to increase the future use and the effectiveness of the SEA application in order to achieve its main goals, laid down in Article 1 of the SEA Directive: "The objective of this Directive is to provide for a high level of protection of the environment and to contribute to the integration of environmental considerations into the preparation and adoption of plans and programs with a view to promoting sustainable development, by ensuring that, in accordance with this Directive, an environmental assessment is carried out of certain plans and programs which are likely to have significant effects on the environment."

These basic recommendations should be adopted:

1. The SEA community should learn from marketing activities and should promote its "product" SEA by demonstrating its potential and usefulness.

2. Specific guidance is necessary on various SEA issues.
3. The SEA Directive is only an intermediate stop on the way to a comprehensive system of environmental assessment.
4. An extension of the Directive's scope to policies, as well as to other areas and/or sectors, and other types of knowledge transfer, should be strengthened.

LAND USE PLANNING AND STRATEGIC ENVIRONMENTAL ASSESSMENT [8]

The concept of SEA has evolved over time from the application of stages of a project-based process to policies, plans and programs to a wider interpretation, often implying an integration of the SEA process into the planning process. Currently SEA may be defined as a formal, systematic and comprehensive process that evaluates the environmental impacts of policies, plans and programs, considers alternatives, includes a written report on the findings of the evaluation and uses the findings in publicly accountable decision making.

Land use has a strong planning tradition, perhaps most directly linked to project level EIA, which gives SEA in land use planning certain specificity. Land use plans SEA addresses the environmental consequences of a land use plan in which the results of a sector policy are translated into physical distribution of certain factors at certain locations.

Land use planning is part of a wider system of spatial planning and perhaps one of the most commonly used world wide types of planning used world-wide – since the very use of space almost automatically implies some form of land use planning. Land use planning reflects the sector policy in a certain area and at a certain level of government, as well as manages different activities and uses of land in a certain area.

Formal requirements for objective led SEA are still rarely encountered in national legislation. However, more or less formal provisions and guidelines for taking environmental aspects into account in land use planning systems are in place in many countries, often dating back before the official adoption of the SEA Directive in 2001. The arrangements for land use planning vary across different countries or sometimes even regions within countries, according to legal, political, economic and cultural conditions.

SEA, and in particularly SEA practice, is still a relatively young discipline. Therefore:

1. There is a need for simple practical guidelines at local level, in particular (but not limited to) on the practical implementation of the SEA Directive.
2. The need for exchange of information, with special stress on practical experience. (Exchange of information is also in

conformity with the requirements of article 12 of the SEA Directive).
3. In many countries, an introduction or amendment of legislation concerning SEA is envisaged; there is a need both for research on effectiveness of certain measures, in particular, types of administrative systems and for the exchange of information on practical experience.
4. It is perceived that currently there is enough knowledge and experience for the preparation of targeted training courses on SEA.
5. Participants in the Pilot Study recommended the areas within SEA which still need most to be developed are quantity and quality of SEA reports, methodology, effectiveness, public involvement, and SEA/EIA relationship.
6. More experience could and should be gained though pilot projects. The use of SEA, including and even stressing, the voluntary application of SEA, should be encouraged as part of activities funded by large international NGO's or financing bodies.

WATER MANAGEMENT AND EIA [6]

The specific interest for environmental impact assessment (EIA) and water management had emerged by the end of the pilot study and as a consequence a in-depth debate has not taken place. However, participants have shown a lot of interest in the complexity of water management and EIA. This pilot study focused on three main themes dealing with water management and assessment of water systems:

1. Integrated water management (IWM)
2. EIA and water projects
3. Transboundary aspects of water management projects and policy.

World-wide, there is a growing tendency for a watershed or river basin approach. The new European Water Framework Directive (Directive 2000/60/EC of the European Parliament and of the Council of 23 October 2000 establishing a framework for community action in the field of water policy, further called WFD)1 and urges a watershed approach, indicating an important step towards an integrated approach for water management.

The new integrated approach for water management is based on several principles, already present in the European environmental policy: high level of protection; precautionary principles; preventive action, damage to be rectified at the source; the principle of the polluter pays; integration of the policy in other community policies; use of available scientific and technical data; variability of environmental conditions of the community; costs/benefits; economic and social development of the community; and international co-operation. The pilot study focused on the kind of problems water management introduces in EIA procedures and how EIA principles and ideas are implemented in water management (WM) projects, plans and policies. The objective was to analyze the implementation and compliance with EIA legislation into WM legislation, especially concerning the integration of the EIA principles; participation of the stakeholders in the WM projects (EIA process for WM projects); common strategies at international, national and regional or local level and the integration of water policy into other policies.

Topics of specific interest and recommendations were:

1. Approaching basins and watersheds as a whole, integrating knowledge in the field of natural sciences, social and political sciences, economical sciences, and considering the legal frames.
2. A better integration between water protection objectives into other policy areas, especially agriculture, infrastructure and land use planning.
3. Transboundary aspects, focusing on the difficulties caused by the difference between geographical borders on one hand and political and administrative borders on the other hand. Water resources can be the cause of instability and even wars, and in more 'stable' conditions would be a source of mutual understanding, and sustainable development for neighbouring regions and countries.
4. Rational protection and use of water systems considering the quality of the ecosystem and respecting self-supporting needs for the functioning of a water system. Rational use includes considering the real price of water.
5. Promoting transparent and participatory water management, one that is not based purely on technical expertise or bureaucratic decisions, but one that takes account of the actual needs of citizens and environment.
6. Promoting the use of financial incentives for better environmental protection and the phase out of perverse subsidies.
7. Establishing criteria to assess a water system, including indicators defining the carrying capacity of water systems. Therefore information on the functioning of water systems is a key element.

TRANSPORTATION AND EIA [7]

The importance of transportation in society is evident when assessed in terms of ecological and geographic features, as well as economical, political, strategic, social and recreation issues. It is also an important aspect in the improvement of quality of life. The economical role is mostly emphasized, due to the understanding that transportation is a competitive advantage factor in the regional or worldwide environment.

Not only did transportation change the modes of mobility of individuals or communities, it promoted new kinds of enterprises as well. Regions and nations were able to integrate, markets develop, technology promoted, economies of scale increased and the economical development of large areas of this planet became a fact. During the nineties, the relevance of transportation increased with the globalization process.

The advent of the "sustainable development" concept (besides its conceptual problem issue) influenced the definition of new perceptions on the environment. This concept, besides dealing with the ecological questions, respects social justice and economical equity values, tries to prevent the social discrimination and to ensure individual, as well as community safety, and respect to human rights in an intra-generational (for the time being) and intergenerational (for the future) perspective.

The traditionally assumed ecological approach is now completed with the emergence of social values, with the enhancement of the social equity issue intending to distribute the costs and benefits in social terms for an effective equilibrium of society. This change in perception also affects the transportation issues.

Recommendations and topics that acquire specific interest are:

1. Three visions on sustainable mobility (high-technology, automobile industry, low activity).
2. Transport and capacity of natural resources.
3. Management opportunities.
4. EIA is increasingly important tool used in transportation development.
5. It is need of SEA in the transportation sector.

CONCLUSIONS

The various pilot study reports contain numerous recommendations on general and specific issues. Broadly speaking, we can formulate some requirements for good practice: "Expertise, public participation, transparency of assessment processes, quality control, review and consideration of the strategic level."

REFERENCES

1. Aschemann, R. 2002. Strategic Environmental Assessment, In: Methodology, Focalization, Evaluation and Scope of Environmental Impact Assessment, Final report, NATO-CCMS Report No.260, p. 89-95
2. EC Contract B4-3040/99/136035/MAR/C2, Environmental Impact Assessment for the Decommissioning of Nuclear Installations, FINAL REPORT, VOLUME 1, 2, 3, Cassiopee, the University of Aberystwyth, ECA Global, 2001
3. Methodology, Focalization, Evaluation and Scope of Environmental Impact Assessment, Final report, NATO-CCMS Report No.260, 2002

4. Pavlickova,K., Bianchi,G., Rosova, K. Public Participation in the Environmental Impact Assessment in the Slovak Republic, Acta Envir.Univ.Comenianae, 4-5, 177-182, 1995
5. Pavlickova,K. Public Participation. 2002. In: Methodology, Focalization, Evaluation and Scope of Environmental Impact Assessment, Final report, NATO-CCMS Report No.260, p. 29-43
6. Pentinat,S.B., Coenen,M., Grishin,N., Kriuaciuniene,J. 2002. Water Management and EIA, In: Methodology, Focalization, Evaluation and Scope of Environmental Impact Assessment, Final report, NATO-CCMS Report No.260, p. 55-69
7. Rosa,M.P., Delcorps,M. 2002. Transportation, In: Methodology, Focalization, Evaluation and Scope of Environmental Impact Assessment, Final report, NATO-CCMS Report No.260, p. 71-87
8. Rzeszot,U.A. 2002. Land Use Planning and SEA, In: Methodology, Focalization, Evaluation and Scope of Environmental Impact Assessment, Final report, NATO-CCMS Report No.260, p. 45-54
9. (9)Van Ghelue, P., Nagels,K. 2002. Quality Control in EIA and Effectiveness of EIA, In: Methodology, Focalization, Evaluation and Scope of Environmental Impact Assessment, Final report, NATO-CCMS Report No.260, p.5-27

PART III:

ENVIRONMENTAL SECURITY

Environmental Management Systems, Reliability Management, and Vulnerability Assessments: Potential within Contemporary Security Settings

John J. Barich, III, and Katarina Mahutova
U.S. Environmental Protection Agency

INTRODUCTION

The U.S. Environmental Protection Agency has defined environmental security as "a process whereby solutions to environmental problems contribute to national security objectives." There are many non-military threats to national security, and several of these are associated with environmental conditions or competition for natural resources. Environmental stress is viewed as deriving from scarcity and resource degradation. It is most easily understood from the perspective of constraints, or "encroachment." Conflict and environmental stress are shown to be related in complex ways. And success in managing conflict is one aspect of overall national security.

Three complementary management tools, the Environmental Management System (EMS,) Vulnerability Assessment/Consequence Management (VA/CM,) and Reliability Management/Failure Analysis (RM/FA) are available to assist environmental managers in assuring that environmental stress as a factor in producing or sustaining conflict is minimized.

DISCUSSION

Environmental Stress and Conflict

Many non-military threats to national security are now recognized. A partial list adapted from NATO includes:

- Environmental mismanagement
- Natural resource depletion
- Overpopulation
- Environmental consequences of the Cold War
- Economic decline
- Territorial disputes
- International terrorism

K. Mahutova et al. (eds.), Defense and the Environment: Effective Scientific Communication, 103–110.

Certain of these threats are directly related to the environment, e.g., environmental mismanagement and environmental consequences of the Cold War. For some, the connection is easily made, e.g., natural resource depletion. For others, the environment may be a factor. How is overpopulation related to the limited ability of environmental resources to sustain human populations? Is economic decline traceable to a lack of markets, a trained workforce, business and financial infrastructure (including capital formation capacity,) or is it traceable in part to environmental resource scarcity? How often are environmental resources a factor in territorial disputes? And is international terrorism a matter for the military and police, or do the environmental threats posed by certain of the tools of terrorists (chemical, biological agents, "dirty bombs,") make terrorism of primary concern to environmental managers?

Most of these non-military threats are caused to some extent by resource constraints. Constraints can be viewed as "encroachment," as illustrated in Figure 1, adapted from the U.S. Air Force:

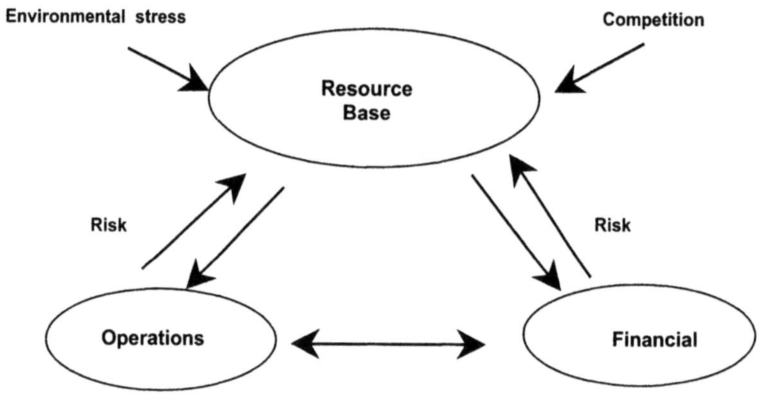

Figure 1. Resource Constraints and Management (Encroachment)

A resource base can be either renewable, non-renewable, or a combination of each. Whereas environmental stress and competition both tend to diminish a renewable or non-renewable resource base, there is some potential to manage environmental stress in a beneficial manner. Operations (such as training) or financial management strategies can either diminish the resource, enhance it (in the case of renewables,) or alter the depletion rate (in the case of non-renewables.) How this occurs is the subject of risk assessment/risk management, i.e., the application of the risk paradigm.

There is a temporal aspect to managing environmental and natural resource systems. This is illustrated in Figure 2, also adapted from the U.S. Air Force:

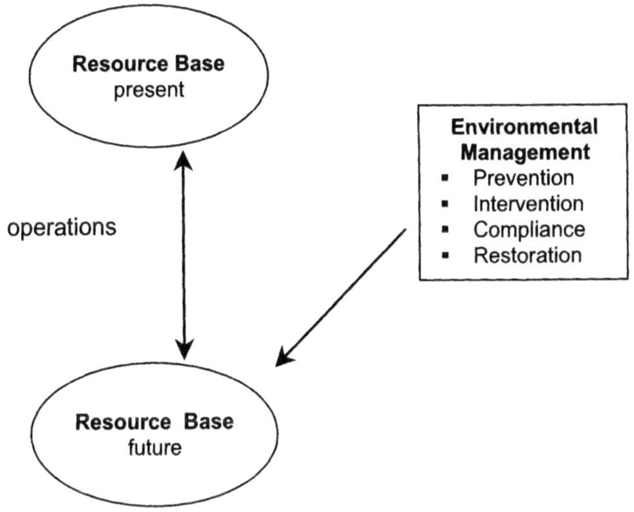

Figure 2. Natural Resource Systems, Present and Future

The quality of the future resource base is very much dependent on effective environmental management during the period of operations. Prevention (including waste avoidance, waste minimization, preference for renewables over non-renewables, and second-use over first-use resources,) interventions to change current practices to enhance the future conditions, compliance with accepted standards and norms, and restoration where past damage has occurred, all work together to assure the best possible future resource base.

Environmental stress is an important factor in the overall quality of the resource base, both now and in the future. This stress has as its source resource scarcity or resource degradation. Environmental stress is closely linked to conflict. NATO uses three phenomena to describe the relationship of environmental stress and conflict. These include:

- multi-modalities
- reciprocity/feedback loops
- consequences of environmental stress

Multi-modalities refer to the observation that environmental stress produces consequences, normally negative, in and through a variety systems, as summarized in Figure 3:

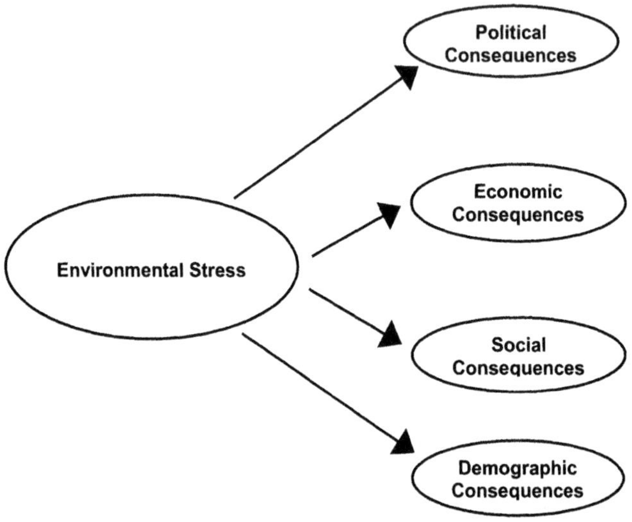

Figure 3. Multi-Modalities: Systems Influenced by Environmental Stress

The environmental manager called upon to evaluate the potential for environmental stress that produces or modifies consequences must have access to multidisciplinary teams including political scientists, economists, social scientists, and demographers. It is not enough to include the 'normal' environmental disciplines of environmental engineering, chemistry, biology, ecology, toxicology, and so forth.

Environmental stress that produces conflict can be influenced by the conflict so as to change the character of the environmental stress. This feedback, or reciprocity loop can be either reinforcing or damping, with reinforcing being more likely. This relationship is depicted in Figure 4:

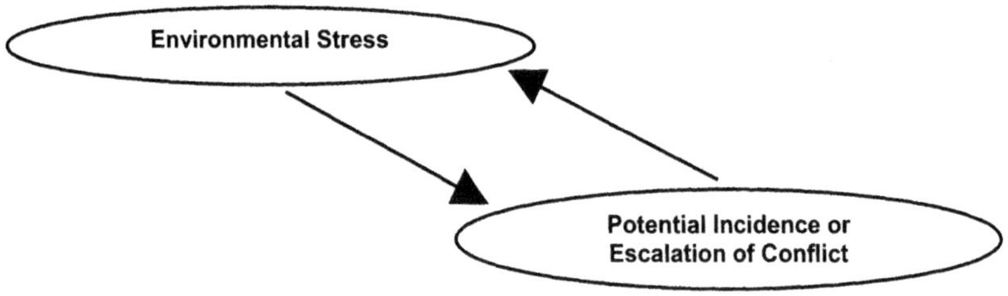

Figure 4. Environmental Stress and Conflict: Reciprocity/Feedback Loops

Finally, as each of the factors of multi-modalities, consequences of environmental stress, and reciprocity/feedback loops are considered together, a complete picture emerges. This is summarized in Figure 5. Of note is the indirect production or modification of conflict through consequences induced in the various modalities, and the feedback loops that exist between conflict and environmental stress.

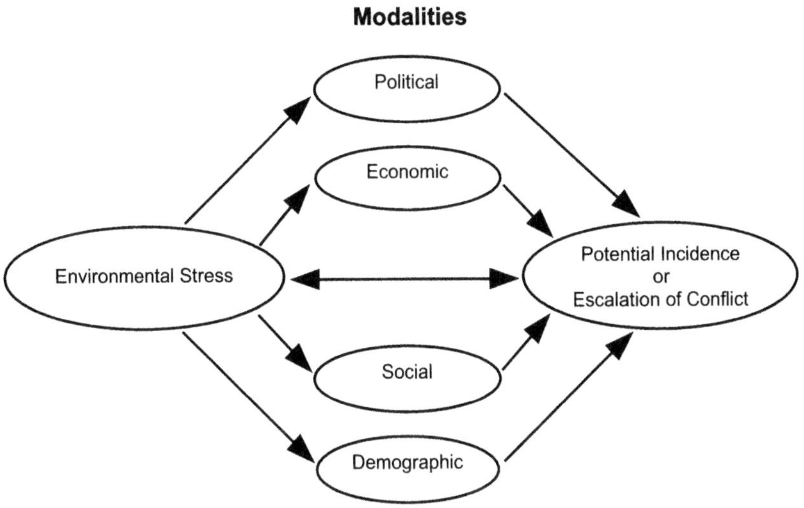

Figure 5. Environmental Stress and Potential Conflict

Environmental Management Tools

Environmental managers have at their disposal three management tools that are useful in responding to environmental security challenges. These include Environmental Management Systems (EMS,) Vulnerability Assessments/Consequence Management (VA/CM,) and Reliability Management/Failure Analysis techniques (RM/FA.) Each relies on a systematic and well-documented approach, and each is widely used in the civilian as well as the military sector.

Environmental Management Systems begin with a commitment from an organization and its management to conduct all operations in accordance with agreed upon environmental policies and procedures. A strength of EMSs is the commitment to continual improvement as realized by a continuous process of plan-do-check-review. Environmental aspects are a key building block for the continual improvement cycle. These are features of any system that might effect the environment, either positively or negatively. Environmental aspects are subjected to impact analysis to help establish priorities for action. If environmental security aspects are included in the EMS inventory of aspects for a facility, the process of continual improvement can be applied to them the product being an environmental security action plan.

Reliability Management/Failure Analysis comprise a set of recognized tools that address the following questions:

- What can do wrong?
- How bad is it or might it be?
- Will planned change yield significant benefits?
- What is the preferred response action?
- How are situations/responses best documented and communicated?

Documentation and communication of situations and responses becomes critical if failures of the past are to be avoided in the future.

Reliability Management normally follows a process of six steps including:

1. Predict the performance of a system with special emphasis on the identification of failure modes.
2. Identify the root causes of a potential or actual failure, and construct event trees leading from the failure to the root cause.
3. Characterize the consequences of failure. This can be accomplished prospectively when attempting to avoid failure, or as an element of response once a failure has occurred.
4. Identify, analyze, and select mitigation (in the case of failure) or avoidance (in the case of normal operations) measures.
5. Implement the mitigation/avoidance measures.
6. Communicate constantly, throughout the process and after mitigation/avoidance measures are complete.

Unlike the EMS, the RM/FA process is less likely to be one of continual improvement, being driven instead by need.

Step 1 recognizes that every system has a weakest link. By systematically identifying these, a list of sub-systems or components are isolated for potential improvement.

Step 2, root cause analysis facilities the analysis of the entire chain-of-events that leads to failure, and provides the opportunity to intervene strategically. In environmental systems, monitoring might be focused on the conditions of the root, thereby providing very early warning of an incipient failure. An example might be detecting contaminants introduced into the vadose zone before they appear in a drinking water supply.

Characterizing the consequences of failure, Step 3, is important in developing bounds for how much to invest in avoidance actions. Every component or sub-system of every system can at a cost be improved from a reliability perspective. Step 3 provides the basis for ranking candidate actions according to benefits to be realized.

Step 4 and Step 5 are steps that are common to the design, construction, and operation of any civil or military system.

Communications, Step 6, are especially important tools of RM/FA. If there is any benefit that comes from failure, it is that a case study is provided that will allow the likelihood of a similar failure to occur in the future to be diminished (but not eliminated.)

Vulnerability Assessments are new tools within the environmental sector. They are becoming important as organizations assume that the threats to environmental security due to willful harmful acts are increasing. VAs begin with a Hazard Evaluation of the environmental system (e.g., a water supply.) The key question is, "What are the vulnerabilities?" A consequence assessment is then completed for each vulnerability. The key question is, "How severe will the consequences be if this vulnerability is exploited?" Next is a mitigation assessment where the requirements of response and recovery are identified. A security assessment is the final step of VA. The key question is, "What is the best strategy to respond to, or upgrade the vulnerable system?"

A review of each of these tools reveals similarities and differences. Whereas EMSs have as their primary focus systems, for example, RM/FA techniques focus on components and sub-systems, while VA/CM tools focus on events. Some of the common characteristics of these tools is presented in Table 1:

Table 1: Attributes of EMS, RM/FA, and VA/CM Systems

	EMS	RM/FA	VA/CM
Management Process	+++	+	++
Planning	++	++	++
Operations	++	+	+
Response	+	+	+++
Feedback	++	+++	+

Each of these tools shares includes a well-defined management process, a commitment to planning, operations or response, and feedback. The management process for the EMS is more developed than for the other two tools. Each is equally proficient at planning. The EMS does the better job of guiding daily operations, while the VA/CM tools do the best job of response and consequence management. Finally, because of its specificity and motivated by the need to repeat past failures, the RM/FA tools provide the best example of the use of feedback.

SUMMARY

Of the many non-military threats to national security, several are associated with environmental conditions or competition for natural resources. Environmental stress derives from scarcity and resource degradation, and is most easily understood from the perspective of constraints, or "encroachment." Conflict and environmental stress are related in complex ways involving multi-modalities, reciprocity and feedback loops. Consequences are expressed through economic, social, demographic, and political systems. Success in managing conflict is one goal of achieving environmental security.

Environmental Management Systems are well-known to environmental managers within both the military and civilian sectors. Their key attribute is a focus on systems and not events. Vulnerability Assessment are relatively new to the environmental sector. Consequence Management has a long history within the environmental sector (response to oil and hazardous chemical spills being typical examples.) Their key attribute is a focus on events and not systems. Reliability Management/Failure Analysis is a set of techniques and tools well-known to the manufacturing and civil infrastructure sectors and are now becoming a recognized tool within the environmental sector. Their key attribute is a focus on the components of complex systems. Together, these three tools can assist environmental managers in achieving environmental security objectives.

Non–ionizing Electromagnetic Radiation and Environmental Security

Andrejs Silins[1] and Ilgonis Klavins[2]
[1] *Latvian Academy of Sciences, Riga, Latvia*
[2] *Defence Property State Agency, Riga, Latvia*

INTRODUCTION

Electromagnetic radiation (ER) is an essential component of the Universe. ER is generated by stars, including Sun, all the time. From these sources, ER with the speed of light, reaches all other objects in the Universe, including Earth. All other objects in the Universe (including Earth) also generate (radiate with the intensity which depends on temperature) and reflect ER. In this way, environment and human beings on the Earth steadily are exposed by ER. Environment and human beings themselves are also generating ER in very small intensities.

Besides these natural sources of ER, humans, in the process of development, have created a lot of artificial sources of ER, starting with fire, radio and television stations, different light sources, X–rays etc. ER is a by-product in other fields of human activities, for example, near electric power lines, and, in fact, near all electrical equipment that uses alternating current.

These natural and artificial sources of ER together create a radiation field which affects the environment and human beings. The intensity of ER in this field is different in different parts of the Earth. Moreover, this intensity is changing many orders of magnitude over time. Therefore, it is necessary to understand the nature of ER and put forward limits (standards) to protect environment and human beings.

COMMON FEATURES OF ELECTROMAGNETIC RADIATION (ER)

When the electric and magnetic fields are time dependent, they influence each other and become coupled [1]. As a consequence of this coupling, electric and magnetic fields can transport energy over much larger distances than the steady electric and magnetic fields themselves. The coupled fields produce traveling waves called electromagnetic waves which propagate through empty space at the speed of light. From Quantum Physics, it follows that these electromagnetic waves (radiation) have particle-like properties.

K. Mahutova et al. (eds.), Defense and the Environment: Effective Scientific Communication, 111–117.

Electromagnetic radiation consist of quanta, each carrying definite amount of energy which is proportional to the wave frequency of the radiation. These quantum properties of ER are essential for understanding the interaction of ER with environment.

Matter can interact with ER by absorbing only definite portions of energy (quanta) of this radiation. When energy quanta is absorbed, then some elements of matter (atoms or molecules) are excited or ionized. Spectrum of energies of excited states are different for different substances. Therefore, any definite substance can interact only with specific quantum energy spectrum, e.g. with ER with specific wave frequencies. For ER with other wave frequencies, this substance is transparent. For example, these properties are widely using in practices, such as manufacturing optical glass for windows.

Real influence of ER on environment and human beings starts when quanta of energy of this radiation are absorbed. Nature of processes in matter, when ER is absorbed, primary depends on the energy of absorbed quanta. When these energies are large enough to ionize atoms or molecules in the matter, then irreversible processes are going on. This is harming environmental security. Therefore, ER is divided in two parts: ionizing ER and non-ionizing ER depending on wave frequencies or quantum energies (Table 1).

Table 1. Characteristics of electromagnetic radiation

Character of radiation	Wave frequencies, Hz	Quantum, energies, eV
A) Ionizing radiation	above $7 \cdot 10^{14}$	above 3
1. Gamma rays	above 10^{20}	above 10^4
2. X – rays	$3 \cdot 10^{17}$ - 10^{20}	$10^2 - 10^4$
3. Ultraviolet radiation	$7 \cdot 10^{14}$ - $3 \cdot 10^{17}$	3 - 100
B) Visible light	$4 \cdot 10^{14} - 7 \cdot 10^{14}$	1,8 – 3,1
C) Non – ionizing radiation	below $4 \cdot 10^{14}$	below 1,8
1. Infrared radiation	$4 \cdot 10^{14}$ - $3 \cdot 10^{11}$	1,8 - 10^{-3}
2. Microwaves	$3 \cdot 10^{11}$ - 10^8	$10^{-3} - 10^{-6}$
3. Radio waves	below 10^8	below 10^{-6}

The border between ionizing and non–ionizing ER depends on the chemical composition of matter where ER is absorbed. Usually the ionization starts when quantum energies of ER are slightly above the quantum energies of visible light. Therefore the ER visible light region is taken as a border between ionizing and non–ionizing ER. In common non–ionizing electromagnetic radiation includes visible light, infrared radiation, microwaves and radio waves.

CHARACTERISTICS OF NON–IONIZING ELECTRO-MAGNETIC RADIATION

Electromagnetic radiation in the visible and near infrared region mainly excites electronic system of the matter. As a result of such excitation, additional chemical reactions could occur which are not possible without the influence of ER. For example, by the influence of visible light, photosynthesis in green plants is going on, which is very important for the life on the Earth. Special materials are developed in which by infrared radiation free electrons are created. Using such materials, unique devices are created to be able to detect warm substances in the night. These devices in our days are efficiently used in military sector for night vision.

By influence of far infrared radiation, microwaves and high frequency radio waves (above 1 MHz) mainly harmonic motions (vibrations) of atoms or molecules in the matter are excited. This usually doesn't stimulate new chemical reactions, but raises the temperature of the matter, because the intensity of atomic (molecular) vibrations determines the temperature of the matter. Therefore by the influence of this type ER on the environment and human beings, the temperature could be increased. Amount of absorbed radiation quanta in the time unit (absorbed radiation intensity) determines the real temperature increase. If the absorbed ER intensities are high, then there could appear reasonable temperature increase in the environment. In this case ER influence could be harmful, because if the temperature, for example, of some living being is raised for some degrees, then it could create pathophysiologicial shifts.

By influence of low frequency radio waves (below 1 MHz) on the environment and human beings, their temperature also could be raised, but in this case, additional effects could be detected. It is caused by different chemical processes in complicated biological systems and some of them involve alternating electric charge transfer. The frequencies of this alternating charge transfer are in the range of low frequency radio waves. If the absorbed ER frequency is equal or close to the chemical process of alternating charge transfer frequency, then by resonance, this chemical process could be stimulated or depressed and the normal process in biological system could be destroyed. In this way low frequency ER could additionally harm the environmental security by the depression or distortion of normal chemical processes in biological systems.

Special attention is paid to the ER with 50 Hz (60Hz in the USA) frequency coming from electric power. This ER is found around all electrical appliances, house wiring, power lines and high voltage transmission lines. There is clear evidence that this ER can produce various hormonal and other changes in living things. It is not yet clear if these changes can result in risk to public health. The results of scientific studies of the possible health risks are very complicated.

Common non-ionizing ER, when absorbed by environment and human beings, are not directly destroying normal processes (as it is doing ionizing ER), but could effect these processes by rising temperature. Strength of influence of non-ionizing ER on environment depends mainly on the intensity of absorbed radiation and on the frequency, especially in the low frequency radio wave region. Therefore special standards (limits) are set up for the intensities of non-ionizing ER to secure environment and human beings.

ENVIRONMENTAL SECURITY STANDARDS

Here we will focus on the influence of high frequency radio wave (above 1 MHz) and microwave ER influence on environment. Absorption of this radiation mainly creates a temperature rise of the absorbing matter. As the temperature changes can reasonably influence the processes in biological systems, the security standards are worked out for human beings.

On that topic, the main work is doing the International Commission on Non–ionizing Radiation Protection. In the recent guidelines [2] the main principles for environmental security standards are discussed. It was determined that in the high frequency radio wave and microwave region the main parameter is the specific absorption rate (SAR). The unit of SAR is watts per kilogram (W/kg). As the penetration depth of ER in the matter depends on frequency of ER (it is lager for lower frequencies), then the SAR is different for ER at different frequencies for the same radiation intensity. Therefore standards for ER intensities are different in different frequency regions.

In guidelines [2] it was established that by influence of ER at frequencies from 10 MHz to 300 GHz, heating is the major effect of absorption of electromagnetic energy, and temperature rises of more than 1 – 2°C can have adverse health effects. It was demonstrated that exposure for up to 30 minutes, under conditions in which whole-body SAR was less than 4W/kg, caused an increase in the body core temperature less than 1°C. Therefore it was concluded, that SAR of at least 4W/kg is needed to produce known adverse health effects in people exposed to ER in this frequency region.

The sensitivity of various types of tissue to thermal damage varies widely, but the threshold for irreversible effects in even the most sensitive tissues is greater than 4 W/kg under normal environmental conditions. These data form the basis for an occupational exposure restriction of 0.4 W/kg, which provides a large margin (ten times) of safety. For general public exposure restriction of 0.08 W/kg is provided, with very large margin of safety (50 times less than threshold). On the base of these restrictions, International Standards for ER intensities which could reach environment and human beings are worked out. In Latvia European Prestandard ENV 50166 – 2 [3] for human exposure to electromagnetic fields in high frequency region (10 kHz to 300 GHz) is introduced as a law. From this Standard it follows that for occupational (worker) continuous exposure the ER intensity in the frequency region 400 – 2000 MHz must be lower than frequency/40[W/m^2] , which for 1 GHz ER gives, the value – 25 W/ m^2 . For general public intensity of such continuous exposure radiation (1 GHz ER) must be lower than 5 W/ m^2.

Most of modern technical equipment is working on pulsed microwaves. It means that radiation from such equipment is coming as short (~ 10^{-6}s) ER pulses and between these pulses is more than 1,000 times longer empty time interval. Also this type of ER raises temperature and that the temperature increase is proportional to the absorbed ER energy in long time. It means that intensity of ER inside short pulses could be reasonably higher (approximately 1,000 times) then for continuous exposure.

On the base of this assumption in the Standards for pulsed exposure of occupational personal the ER intensity in the frequency region 400 – 2,000 MHz must be lower than frequency multiplied by 20.4 , which for 1 GHz ER gives the value equal to 20,400 W/m^2 . For general public, intensity of such pulsed exposure radiation (1 GHz ER) must be lower than 3,970 W/ m^2 . As we see these values are much higher (almost 1,000 times) than standardized ER intensity limits for continuous exposure. But, in pulsed the ER case, we must be careful if the pulse sequence are in the region of low frequency radio waves. Then additional effects on human beings could appear.

Compared with continuous wave ER, the pulsed radiation with the same average rate of energy absorption in tissues are generally more effective in producing biological response [2]. The "microwave hearing" effect is a well known example. People with normal hearing can perceive pulse-modulate ER with frequencies between about 200 MHz to 6.5 GHz. People are variously describing this effect as a buzzing, clicking or popping sound, depending on the modulation characteristics of the ER. The microwave hearing effect have been attributed to thermo elastic interaction in the auditory cortex of the brain.

MODERN TECHNICAL EQUIPMENT IN THE MILITARY SECTOR

Radar is one of the most widely used modern technical equipment in the military sector. It is mainly used for the air surveillance. Starting from 1996 Latvia, together with other Baltic States, are working on common air surveillance system – BALTNET. According to the plans of this system in Latvia in 2003, a new long range solid state radar TPS–117 will be installed. In 2002, wide public discussion began on that subject connected with environmental and people security aspects. The Ministry of Defense of the Republic of Latvia asked the Latvian Academy of Sciences to perform special investigation on possible radar TPS–117 influence on environment. A working group of scientists from different fields of sciences was created and worked together for several months. Deep analysis of possible radar TPS–117 ER influences on the environment was done and reviewed to the Ministry.

Special attention was paid to the investigation of the nature of radar TPS – 117 ER. Taking into account technical parameters provided by the Lockheed Martin company, theoretical calculations of radar pencil beam ER intensities and other parameters at several distances from radar were performed. Results are shown in Table 2.

Table 2. Calculated parameters of radar ER

Distance from radar, m	ER intensity W/m^2	Magnetic field strength, A/m	Electrical field strength, V/m
85	1510	2,001	754
100	1095	1,704	643
125	700	1,363	514
170	279	1,003	378
200	274	0,852	321
260	162	0,656	247
520	41	0,330	124
1000	11	0,052	19

Besides theoretical analyzes these parameters were calculated by engineering methods. As the engineering calculations take into account the power losses between transmitter and antenna, which is 1.3 times, then the obtained values for radar ER parameters are 1.3 times smaller than theoretical ones. In fact, it shows that both methods are giving the same results, because power losses between transmitter and antenna must be taken into account in theoretical calculations.

Engineering method allows estimating the radar ER intensity close to the antenna (below 38m) and that is 25.20 W/ m^2. Besides that the ER intensities apart from the pencil beam were estimated. If the antenna is situated 12 m from the ground level, then ER side intensities at ground level are: 100m from antenna --- 14 W/ m^2 ; 500m from antenna – 0.5 W/ m^2 and 1,000m from antenna – 0.15 W/ m^2. Experimental measurement performed at the distance approximately 500m from antenna showed ER intensity only 0.14 W/ m^2. As these results are reasonably lower than estimations, they could be taken as the upper limit of side ER intensities.

Standard limits for TPS-117 ER radiation are shown in Table 3.

Table 3. Standards for radar TPS-117 radiation.

Frequency, MHz	ER intensity for workers, W/m^2	ER intensity for general public, W/m^2
1215	24786	4824
1300	26520	5161
1400	28560	5558

When these experimental results and theoretical estimations of radar TSP – 117 ER intensities are compared with strongest standard limits which are 24,786 W/ m^2 for occupational personal and 4,824 W/ m^2 for general public, we could see that real ER intensities coming from radar TPS–117 is essentially lower inside the pencil beam. Outside the beam, on the ground level close to the radar (100m), the ER intensity is more than 300 times lower than strongest Standards value permits. This means that modern technical equipment in the military sector (radar TSP–117) are design in such a way that intensity of ER reaching environment and people is essentially below the strongest limited values of Standards.

CONCLUSIONS

1. Electromagnetic radiation is an essential component of the Universe and plays important role in our every day life.
2. Absorption of the non-ionizing electromagnetic radiation from the microwave and high frequency radio wave region by environment and people is raising the temperature of the absorbing substances.
3. International standards are developed to protect people from adverse health effects and modern technical equipment in the military sector is design strongly taking into account these standards.

REFERENCES

1. Paul M. Fishblance, Stephen Gasiorowicz, Stephen T. Thornton, Physics for Scientists and Engineers. Prentice Hall Upper Saddle River, New Jersey, 1993.
2. ICNIRP Guidelines, Retriewed march 2001 from: http://www.icnirp.de/frman.htm
3. European Prestandard, ENV 50166- 2. CENELEC, 1995

Consequences of the Chernobyl Catastrophe as a Prototype of Nuclear Terrorism

Dmitry M.Grodzinsky
Institute of Cell Biology and Genetic Engineering, National Academy of Sciences of Ukraine, Kiev, Ukraine,

The territory of the Alienated Zone around the Chernobyl Nuclear Power Station is the most contaminated ecosystem. Here the main radionuclides ^{137}Cs, ^{90}Sr as well as transuranic elements are widely distributed. An important point is that the contaminated plots are characterized by very different environmental parameters. Namely, here various soil types, landscapes, subsoil rocks, depths of underground water are distributed throughout the contaminated territory. Thus we stand a good chance of getting information on the biogeochemistry of radionuclide and radiobiological effects of irradiation.

During sixteen years serious efforts were mounted to develop reliable countermeasures as applied to various situations in the territories around the Chernobyl Nuclear Power Station.

The bulk of evidence collected by us favors this view. Because of this consideration of the consequences of the Accident at the Chernobyl Nuclear Power Station in 1986 in respect to the analogies of nuclear terrorism.

The enormity of the September 11th, 2001has shown that mankind should envisage all conceivable forms of terrorist tools.

This paper is devoted to the problems of nuclear terrorism. There are at least two perils of using of radioactive materials as the invading agents.

The first one is a disruption of a reactor core followed by the ejection of huge activity of fission products in environment. The result of core disruption is contamination of soils, natural waters, sources of water, plants, surfaces of buildings and so forth. The consequences of this terrorism event could be very dangerous the health of people so also will have a severe impact on economy of wide regions.

The second one is the spraying of radioactive materials in the environment. Here, we are dealing with the use of a "dirty bomb". These bombs are defined as an explosive device by means of which the radioactive materials are dissipated in environment.

The temperature rise, moving and compression of the products of the explosion, fail in compression as well as a range of material dispersion, by and large, depend on the type of the explosion device and its power.

K. Mahutova et al. (eds.), Defense and the Environment: Effective Scientific Communication, 119–137.

Bearing in mind an illogicality of the evil of terrorism, we have to consider different possibilities of the terrorist action involving a great variety of bugs. We are now in a position to consider different types of the dirty bomb as to the radioactive materials present in explosion device. The most likely fission products are used for this purpose. However, use of individual radionuclides in a dirty bomb must not be ruled out. Tritium, radiocarbon as well as emitters of alpha particles are free to choosing reasons of difficulties for detecting. The fission products are detectable very simply in contrast to the emitters of alpha or beta particles without attendant gamma radiation. It should be remembered that the weight of a large unit of activity for the pure short-lived radionuclides is very neglected. However, the production of pure radionuclides without carriers is normally a complicated and expensive process. Nevertheless, we should consider disparate systems of dirty bombs no matter how wild looking they are.

From the aforesaid it might be assumed that the main reason of the striking effects is radionuclide contamination of the territories with formation of a radiation field.

The Accident at the Chernobyl Nuclear Power Station in 1986 might be considered as a huge and spontaneous experiment on the biosphere as a whole when large activities of different radionuclides were released into the environment. What's more we are be able to find either of the precedents characterized to the consequences of the atomic terrorism as they may be on the intensity of explosion or the types of the explosive substances. The bulk of evidence collected by us favors this view that consideration of the consequences of the Accident can be useful as an analogy to nuclear terrorism.

Radionuclides in the form of ions, carbides, oxides or fuel aerosol particles were propagated with very high velocity into the atmosphere. Hence, the radioactivity of Chernobyl origin in the first days of May was detected everywhere in the Northen Hemisphere: in the Countries of Europe, in the Japan Sea and United States of America.

The estimations of the total activity released from the wrecked reactor varied significantly with time. The last assessment of the spillage of fission products add up to 1.2 10^{19} Bq including about 7 10^{18} Bq activity of the noble gases. About 3 % of nuclear fuel contained in the core at the Accident was released as radioactive materials as well as 100 % of inert gases and 20 – 40 % of volatile radionuclides.

The radionuclide composition of the released radioactive material was very complicated being consistent with the constitution of the old nuclear fuel at the destroyed reactor. Isotopes of iodine with short periods of half life, radionuclides of cesium, strontium, cerium and transuranic elements are the most important substances in the view of radiological effects. The estimations of the activity of these radionuclides contained in releasing are as follows: for ^{131}J – 1.3 – 1.8 10^{18} Bq, ^{134}Cs – 0.05 10^{18} Bq, ^{137}Cs – 0.09 10^{18} Bq. These activities correspond to 50 – 60 % of ^{131}J and about 20 – 60 radioactive isotopes of cesium which were contained in the core. The nuclide composition of the released fission products is shown in the *Table 1*. Contamination of the territories is shown in the *Table 2*.

Table 1. Released radionuclide composition as a result of the Chernobyl accident

Radionuclide	Released activity, Mci		The activity released by 06.05.86, %
	26.04.86	06.05.86	
^{133}Xe	5.0	45.0	Probably, up to 100
^{85m}Kr	0.15	-	The same
^{85}Kr	-	0.9	The same
^{131}J	4.5	7.3	20.0
^{132}Te	4.0	1.3	15.0
^{134}Cs	0.15	0.5	10.0
^{137}Cs	0.3	1.0	13.0
^{99}Mo	0.45	3.0	2.3
^{95}Zr	0.45	3.8	3.2
^{103}Ru	0.6	3.2	2.9
^{106}Ru	0.2	1.6	2.9
^{140}Ba	0.5	4.3	5.6
^{141}Ce	0.4	2.8	2.3
^{144}Ce	0.45	2.4	2.8
^{89}Sr	0.25	2.2	4.0
^{90}Sr	0.015	0.22	4.0
^{238}Pu	$1x10^{-4}$	$8x10^{-4}$	3.0
^{239}Pu	$1x10^{-4}$	$7x10^{-4}$	3.0
^{240}Pu	$2x10^{-4}$	$1x10^{-3}$	3.0
^{241}Pu	0.02	0.14	3.0
^{242}Pu	$3x10^{-7}$	$2x10^{-6}$	3.0
^{242}Cm	$3x10^{-3}$	$2x10^{-2}$	3.0
^{239}Np	2.7	1.2	3.2

Table 2. Areas contaminated with ^{137} Cs In Ukraine Km^2.

Level of contamination, kBq/m^2			
37 - 185	185 - 555	555 - 1480	>1480
37,205	3,177	882	571

There are plots characterized by very different surface radionuclide contamination within this territory. We can find parcels where the surface contamination is over 10^8 Bq per m^2 or range up to 10^3 Bq per m^2. This is the situation in the contaminated territory. The wide variation in surface contamination is caused by peculiarities of radioactive precipitation in which local intensities of rains and wind determine the spotted picture of radionuclide contamination. The sizes of these spot of high radioactivity are variable from some meters to tens of kilometers in diameter. The overall picture of radioactive contamination of the Ukrainian territory with radiocesium, radiostrontium and transuranic elements can be seen in *Figures 1 - 3*.

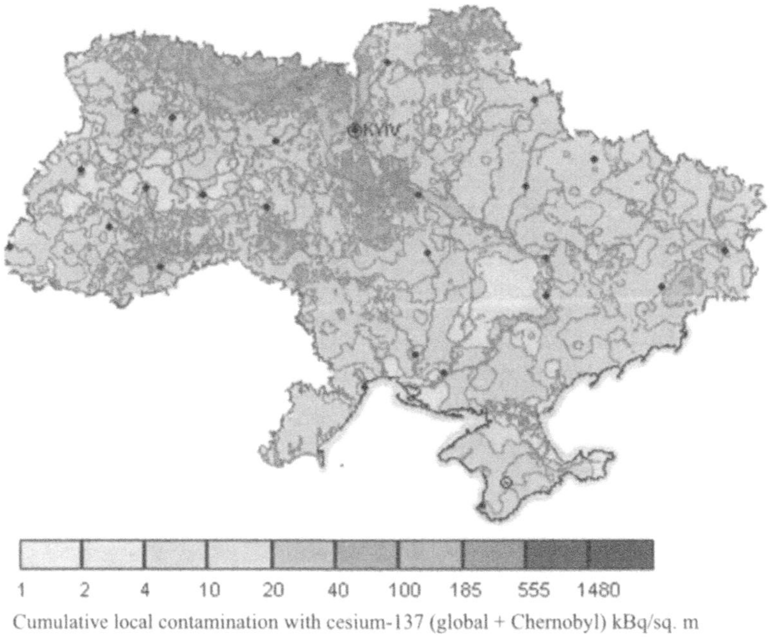

Figure 1. Cesium-137 contamination of Ukraine as of January 1, 1998.

The territory of the Alienated Zone around the Chernobyl Nuclear Power Station is the most contaminated ecosystem. Here the main radionuclides ^{137}Cs, ^{90}Sr as well as transuranic elements are distributed in large activities. The Alienated zone is a peculiar kind of depot from which the dangerous long living radionuclides can transfer to the surrounding territories.

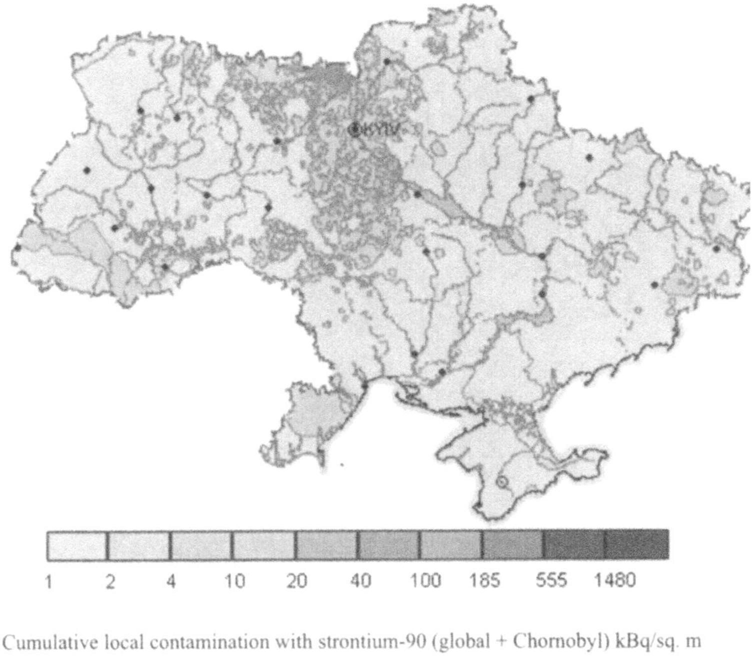

Figure 2. Strontium-90 contamination of Ukraine as of January 1, 1998.

An important point is that the contaminated plots are characterized by very different parameters of the environment. Namely, here various soil types, landscapes, subsoil rocks, depths of underground water are distributed throughout the contaminated territory.

Thus we stand a good chance of getting information on the biogeochemistry of radionuclides and radiobiological effects of irradiation.

Two types of Hazard which are due to the Radionuclide contamination of the region have a profound impact in the life of people. In the first place the state of health of people depends on the dose rate and accumulated dose. The deterioration of health will be the main consequence of the explosion of the dirty bomb or releasing of radionuclides from the wrecked reactor unit. The risk resulting from irradiation can be calculated using risk coefficients deduced from epidemiological data after the atomic bombing of Hiroshima and Nagasaki or the Accident at the Chernobyl Nuclear Power Station. An additional hazard that must be scrutinized a psychological effects from the introduction of radioactivity in the environment. It should be pointed out that psychological stress in the Ukrainian population after the Accident at the Chernobyl Nuclear Power Plant was very strong and a term "radiophoby" has evolved from the evidence of psychological observations.

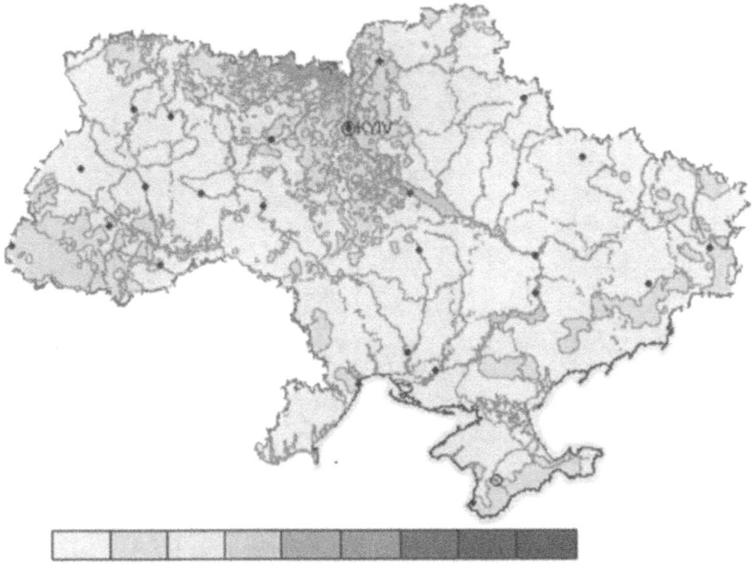

Figure 3. Plutonium isotope contamination of Ukraine

In the second place the hazard for ecosystems should be separately analyzed. Loss of biodiversity, accumulation of radionuclides in individual components of ecosystems can be responsible for the entering of radionuclides into the human-being. By way of illustration we can define the term radionuclide capacity of ecosystem. This term means the value of collective dose for people associated with the certain ecosystem with the constraint of loading this ecosystem with a fixed quantity of radionuclides.

To minimize the ecological hazard from landscapes contaminated with radionuclides it is necessary to make a complete and detailed inventory and assessment of their chemical and physical conditions, and the potential risk to ecosystems. The countermeasures should be employed thereafter.

The main radioecological consequences of the explosion of a dirty bomb or the destruction of the core might be seen from data obtained after the Accident at the Chernobyl Nuclear Power Station. These consequences are as follows:

1. The large activity of fission products escaped into the atmosphere and further into ecosystems. The radionuclides are migrating through all components of the ecosystems in the landscapes and, as a result, all

living beings became radioactive, among others microorganisms, fungi, plants, insects, other animals and, unfortunately, human beings.

2. Radioactivity is moving to ground water and contaminating surface water. Thus drinking water became radioactively contaminated.
3. Radionuclides come to the trophic chains in ecosystem and through these chains are moving to feed products and fodder and to people. All things around the human being become radioactive as well as the bodies of adults and children. For one example the radioactivity of leaves of some trees in the central Kiev's streets in 1986 varied from 70 to 4000 kBq / kg.
4. Dispersed radionuclides migrate in the biosphere and this process is accompanied by shaping of dose load on all living beings including a large number of people. Exposure to population due to the fallout from the Chernobyl reactor accident deposited on the ground via three pathways: Firstly, external irradiation from radioactive materials deposited on the various objects surrounding the people. Secondly, inhalation of airborne materials that are aerosol particles. Thirdly, ingestion of contaminated foodstuff and water. Contribution of ingestion of contaminated foods to the total irradiation dose has been very high. Contribution of the outer irradiation or inhalation of radionuclides might be particularly substantial in the first moments after explosion of a dirty bomb. It is necessary to note that internal irradiation is characterized by its much more high biological effectiveness than external irradiation.
5. Excess of irradiation over the natural background could lead to manifestations of different kinds of diseases in people and deterioration in flora and fauna

Any radionuclide is subjected to cycling when it is becoming trapped in the ecosystem . In the general case the cycling of Radionuclide has the aspect shown in *Figure 4.*

Originally the radionuclide settles on the surfaces of objects in the ecosystem, namely, the surface of soil, plants, water-table, etc. Next is the penetration of radionuclide into soil solution in the forms that are available for plants and microorganisms. The radionuclide bulk of the soil can be partially transformed in a form non-available for plants. Soil absorbing complex is responsible for the bounding of the ionic forms of the radionuclide. This process is partially reversible. Plants absorb the radionuclide together with according carriers. The absorbed by plants radionuclides are trapped further to tissue of phytophages and are moving through the trophic chain to the human beings. The dying parts of plants as well as the remains of animals find its way into the pool of detritus in soil.

The process of circulation of radionuclides is usually slow. That is the reason why the main masses of radionuclide are grouped on the upper layers of soil.

The mobility of radionuclide is characterised by the transfer coefficient. The meanings of the transfer coefficients are varied over wide limits. The greater is the transfer coefficient the more is the mobility of the radionuclide and the more is danger for people. The transfer coefficients are dependent on the plant species and types of soil. Lichens and mushrooms, as a rule, demonstrate the highest values of the transfer coefficients.

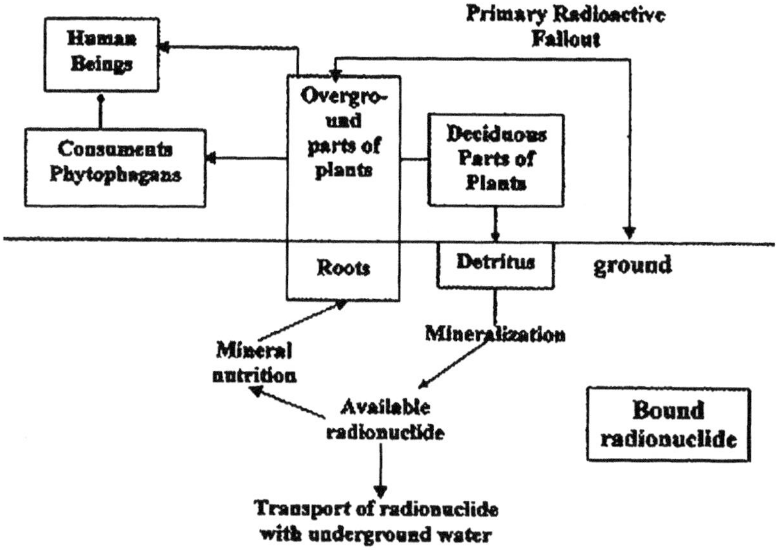

Figure 4: Scheme of radionuclide cycling in terrestrial ecosystems

There are two main strategies of reduction radionuclide absorption by the human organism. The first possibility is that the meanings of the transfer coefficients are restricted by the assortment of the special forms of plants.The second possibility is that the processes of radionuclide bonding in non reversible absorption on soil minerals form is intensified. Application of organic fertilizers, strictly speaking, serves this purpose.

As soon as a dirty bomb has been exploded it is necessary to form a true notion about the spatial distribution of the radioactive contamination. The program of monitoring is to obtain exhaustive information about the content of short-lived and long-lived radionuclides in all components of the environment, namely, in soil, air, water in reservoirs of different kind, plants, on the surfaces of the building (walls, roofs), indoor and outdoor. This monitoring should embraces not only the contents of radionuclides but also their physical and chemical states. Among other things the aerosol dispersion should be evaluated as well as the availability of the hot particles. The monitoring should be organized promptly in the wake of the explosion. From this monitoring starting point it is necessary to keep track of the control of radionuclide transferring by wind, water and biotic components.

Three general types of environmental monitoring are distinguished, namely, basic (standard) monitoring, emergency (rapid) monitoring and scientific (precision) monitoring.

We have an experience monitoring after the Accident at the Chernobyl Nuclear Power Station. The first stage of operations were carried out following different procedures, with different techniques and with different sampling schemes, so the data were almost incomparable. It is apparent that care should be taken to form a unified system of adequate sampling and recording of radioactive contamination and migration. The control of changes in the health for residents of the contaminated areas should be also of paramount importance in monitoring. It is very significant that changes in the health of people are a slow process. This raises the question as to using different biotest system peculiarly designed for detecting the fast radiobiological effects. By way of examples some plants constructed by gene transferring have been applied validly to the assessment of biological effects due to irradiation.

The main goals of radioenvironmental monitoring are as follows:

- control over the state of the zone contaminated with radionuclides, its individual, the most dangerous parts and the measures to reduce their danger;
- monitoring of the natural objects by the same parameters which characterize the environmental situation both in the contaminated zone and beyond its limits;
- detection of tendencies to changes in the environmentally dangerous objects which determine the migration of radionuclides;
- control of tendencies to changes in the health for residents affected by the explosion.

It might be well to create data bases for prediction of the environmental situation in the contaminated zone.

The objects of radioenvironmental monitoring are as follows:

- landscapes and geological medium;
- surface and subsurface aqueous systems;
- ground water, notably the sources of drinking water;
- surface of soils and upper layers of soil;
- air;
- cultured plants (first and foremost, vegetables and fodder plants) and mushrooms;
- milk, meat, fish;
- biocoenoses and its components.

The environmental monitoring is carried out at different scale research levels. The scale of the monitoring sufficiently depends on the intensity of the explosion.

The major components of the Radioenvironmental Monitoring should embrace radiogeochemical processes, behavior of the dissolved radionuclides in natural water, absorption of radionuclides by plants, solubility of hot particles, contamination of food products. These combined data of monitoring are necessary for calculations of the dose load on people and prediction of feasible consequences of the Explosion.

It seems that we should be very attentive to revealing the consequences of the Accident because it is related to the principally new aspects of radiation accidents ecologically dangerous in a large scale as follows:

Firstly, the conditions of combined irradiation by the mixture of beta-, gamma and alpha rays are of rather higher radiobiological effectiveness.

Secondly, effectiveness of the internal irradiation related to the radionuclides deposited within cells and tissue is much higher as compared with the irradiation of one type. The example of this phenomenon might be seen in the *Table 3*, where the dose curves for internal and external irradiation are sketched.

Table 3. Frequencies of waxy reversions in barley pollen grains as a result of 55 days of exposure to various type of irradiation

Dose rate $\mu Sv.h^{-1}$	Total Dose MSv	Total reversion frequency per 10^6 pollen grains	Radiation-induced reversions per 10^6 pollen grains
Conditions of radionuclide contamination			
Control (0.96)	1.3	174	0
59	78	226	52
320	422	837	663
400	528	1235	1061
515	680	1705	1531
Chronic irradiation in gamma field			
Background (0.11)	0.1	82	0
5	3.0	145	63
50	29.6	150	68
500	296	198	116
5000	2960	192	110
50000	29600	292	210

Thirdly, there are synergetic effects. The problem, in essence, is this: the harmful action of irradiation in the presence of some agents of chemical nature can be much higher as opposed to the action of the sole irradiation.

It seems likely that we have no basis to apply the traditional radiobiological knowledge to elucidating, quantitatively evaluating and predicting the real threat of the Chernobyl radionuclide contamination.

The victims of the Chernobyl Accident in official documents are divided into four groups in the following way:

- The first group encloses the participants in the liquidation of the Chernobyl Accident consequences. 245,587 persons fall in this group. Among the participants there are 223,908 men and 21,679 women.
- The second group embraces the people evacuated from the Alienated Zone and emigrants from the zone of obligatory settling out. This group consists of 70,483 persons, among whom 31,365 are men and 39,128 are women.
- The third group covers the residents who are living now or were living some years after the Accident in the Zone with strict radiation control. This group is very numerous, it consists of 2,096,000 persons (45,7 % men and 54,3 % women).
- The fourth group comprises the children born from the parents of any one from the listed groups. This group exceeds 450 thousand persons.

The epidemiological data testify that the morbidity of the victims of the Chernobyl Accident as a whole and by main classes of diseases is higher than the average values in Ukraine and has a very clearly defined tendency of increasing with time. It can be seen from *Figure 5* that the index of morbidity is drastically grown with time.

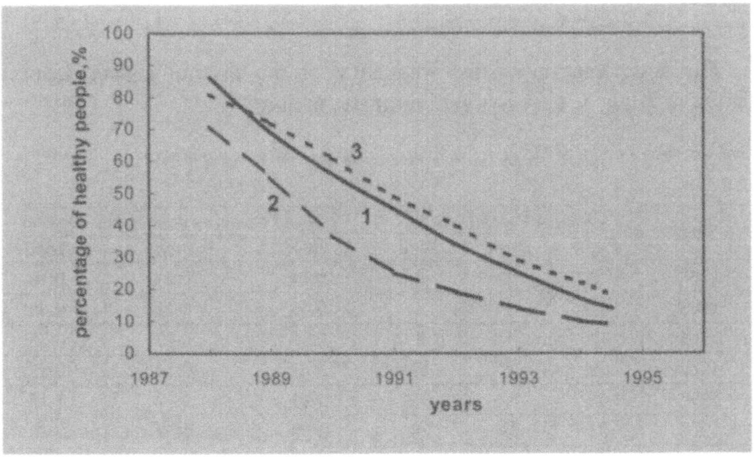

Figure 5. Percentage of healthy people in cohorts of liquidators (1), evacuated from the Alienated Zone (2), and children of irradiated parents (3)

The following syndromes have been established in the initial stage of the Accident:

1. Hemopathology – cytopenia, leukocytopenia;
2. Astheno-vegetative syndrome;
3. Neurocirculatory dystonia;

4. Syndrome of laryngitis;
5. Beta-injury of the mucosa and skin (it was named as the "beta-burn").

The wide variety of nosological forms has been revealed at later years. The structure of the main diseases in the sufferers of the Chernobyl Accident is shown in *Table 4.*

Table 4. Augmentation of morbidity in cohort of the participants of the liquidation of the accident consequences and in adult population of Ukraine (relationships between morbidity in 1988 and 1998).

Nosologic forms	Participants of the liquidations of the accident consequences	Adult population of Ukraine
All diseases	12.40	1.27
Infectious diseases	4.40	1.25
Diseases of endocrinologic system	192.111	1.49
Blood and hemopoietic organs diseases	79.00	3.82
Psychic disorders	13.35	1.00
Nervous system and sense organs diseases	37.31	1.00
Blood circulation organs diseases	112.46	2.20
Respiratory organs diseases	2.55	0.80
Digestive organs diseases	81.28	1.47
Genito-urinary diseases	41.69	1.46
Skin and subcutaneous tissue diseases	2.30	1.13
Osteomuscular system deseases	12.85	1.32

The augmentation of the frequency of the thyroid cancer cases is shown in *Table 5*. This is a very dramatic picture.

Table 5. Thyroid cancer incidence among Belarusian children population.

Year	Cases of thyroid cancer			Thyroid cancer incidence per 100,000		
	Girls	Boys	Total	Girls	Boys	Total
1986	2	1	3	0.14	0.07	0.10
1987	10	3	3	0.71	0.21	0.45
1988	4	4	8	0.28	0.27	0.25
1989	6	5	11	0.43	0.34	0.38
1990	16	17	33	1.13	1.17	1.15
1991	55	23	78	3.90	1.58	2.70
1992	50	41	91	3.54	2.82	3.17
1993	94	33	127	6.66	2.27	4.42
1994	103	46	149	7.30	3.16	5.20
1995	95	49	144	6.73	3.37	5.00
1996	90	43	133	3.38	1.65	2.50
1986 -96	525	265	790	3.38	1.65	2.50

Currently some evidence has been advanced which suggested that the frequencies of the hematological malignant diseases such as acute lymphoblastic leukemia, chronic lymphocytic leukemia of B-cell type, multiple mieloma non-Hodgkin's malignant lymphoma tends to increase

Genetic monitoring is very beneficial in revealing irradiation harm in people. Taking into account the data on the embrional lethality, frequencies of aberrant and polyploid cells, frequency of cytogenetic injures, micronuclei formation, abnormal sperm heads, frequency of translocation, dicentrics and rings of chromosome formation it is possible to develop on understanding of the development of late effects of irradiation. A wealth of experience in the genetic monitoring of sufferers of the Chernobyl Accident has been gained. These observations show uniquely the impact of irradiation in low doses on the genetic apparatus of cells.

I would like to show a sketchy description of the main factors of the radiobiological effects being formed in the territories contaminated with radionuclides.

Two general types of radiobiological effects are recognized: non-stochastic or deterministic and stochastic effects.

Non-stochastic effects consist of somatic damages and deficiencies of the immune system. These effects induce the secondary diseases, which can be manifested as illnesses both specific for irradiation as the radiation disease and sometimes non-specific secondary disease. Manifestation of these effects has a dose threshold. The level of the threshold of radiation sickness in human being ranges up to 0.8 Gray. The intensity of the manifestation of these effects depends strictly on dose, dose-rate, and types of radiation.

Stochastic effects are characterized by the probabilistic nature of its manifestation. These effects have no dose threshold and the intensity of its manifestation does not depend on the dose. Cancerogenesis based on cell transformation as well as mutagenesis of somatic and reproductive cells are referred to as the stochastic effects of irradiation.

The effects of low dose irradiation should be mainly stochastic in its nature. Hence the problem of the stochastic effects should attract considerable attention.

The harm for people can be characterized by different parameters, for instance general morbidity, reduction of average lifetime, frequency of diseases of different nosological forms or by common parameters of the quality of life, which pay off as economic amounts - sum of expenses necessary for an avoidance of effects, and all of the factors negatively reacting on the man.

In addition individual harm it is necessary to consider and evaluate population damages. The quantitative characteristics of the tendencies in changes of a demographic situation are used in this case. Very many factors are known which do not render an essential individual harm, but put the large damage of a demographic situation in region.

There appears to be the most adequate method of attack a problem of risk assessment an approach based on the results of fundamental radiobiological investigations.

We take as the starting point, the derivation of the tree-of-risks for the stochastic effects of irradiation. This tree-of-risks is visualizes as a graph in which sequences of stochastic unitary radiobiological events have been taken into account.

The tree-of-risks displays the pathways leading to harmful cell effects that are well-represented as a consequences of stochastic events. The pathways of radiobiological injury begin with the events of hitting into DNA molecules. The probability of this hitting into the target varies proportionally with the relative size of chromosomes in the cell. The value of this probability can be calculated using equation of Poisson distribution.

The events of hitting a target do not result in molecular damage to molecules.The probability of this type of realization is dependent on the state of chromatin, availability of native or artificial radioprotectors and radiosensitizers as well as some other causes. However, the probability of this event is kept constant for a given cell.

The type of cell, its radiological prehistory and age governs the probability of DNA reparation. In the general case the probability of completeness of DNA reparation should be experimentally estimated. In the case when DNA reparation has been completely effected no harmful consequences in the cell ontogenesis will be manifested as the stochastic effect of irradiation. The fate of the cell will be different for somatic cell and for stem cells or generative cells and depends on species of the organism. In the majority of instances, fate can be well of revealed on the basis of the histological pattern. We can see when an emergence of non-reparable DNA damages or products of error-prone reparations take place. But it should be emphasized that initial value of DNA damages is subject to wide variation in connection with distinctions of starting contamination the cell with of radionuclides. In this connection an evaluation of specific radioactivity of different tissues should be estimated.

The fate of generative cells depends on the efficiency of cell selection. Two types of cell selection are differentiated, namely haplontic and diplontic. Owing to haplontic selection on the whole population of gametes can be cleared of the lethal mutations. In such a manner the lethal mutations will be fully eliminated and only viable mutants and undamaged generative cells retained in the cell population of gametes.

Gametes that are carriers of mutations are responsible for the late radiobiological effects in the form of genetic alterations. This type of stochastic effects of irradiation can be realized during many generations, for example sterility.

The probability of mutants formation in gametes should be experimentally detected. There is a definite numerical relationship among the mutants of different genes, hence the name spectrum of mutations.

The frequency of mutants arising depends on the size of the irradiated population. The greater is population the greater is the yield of mutant forms.

A proliferate death has its origin in chromosome aberrations. This harmful stochastic effect can be balanced by a repopulation in cell population when G_0 cells can be engaged in recovery of functioning stem cells.

The transformation of a stem cell or a somatic cell is the governing factor of cancer formation. Transformed cells are responsible for some forms of cancer under certain circumstances. Among other things, a prerequisite for the formation of cancer is the presence of a definite number of the transformed cells and the failure of the immune system of the organism to recognize transformed cells. The time it takes for accumulation of the appropriate number of transformed cells and formation conditions adequate to tumor growth vary widely for different types of tissue. More than 40 years has elapsed after the Atom bombardments of Hiroshima and Nagasaki, but radiation induced cancerogenesis continues.

The probability of radiation-induced cancer can be calculated on theoretical grounds, namely, efficiency of diplontic cell selection, and the rate of accumulation of transformed cells. As long as the transformed cells are available in tissue the risk of radiation induced cancer exists. The value of this risk can be defined taking the product of the numerical values of the foregoing probabilities or from experimental observation.

Radiation-induced mutations, which emerge in somatic cells, can be often masked by the normal functioning of other cells in low dose cases.

Thus, the assessment of the quantitative values of risks of harmful effects of irradiation can be attained by taking into account elementary events shown on tree-of-risks.

The experimentally accessible methods of an evaluation of numerical significance of probability of the main elementary events in the process of radiation-induced risk of formation are defined. These methods have their basis in utilization of special biotest systems.

The general view of the tree-of-risks depends on the nature of the targets to which the damaging agents are primarily directed. DNA in cell nucleus or mitochondria DNA as well as the membrane systems serve as targets for ionizing radiation. Other cell systems can play a role as targets in the case of influences of alternative factors. In *figures 6-8* these targets and possible effects as a response to radiation stresses are shown.

Target 1 **- GENOME OF CELL NUCLEUS**

The sizes of this target (volume of chromosome}
vary in the range from 10^{-6} to 10^{4} micron^{3}
The yields of the molecular injures in one median cell
Of animal may amount on 1 Gy:
40 double strand break 1000 single strand break
700 8-hydroxyadenine formation 250 damages of thymine

Target 2 **MEMBRANE CELL SYSTEM**

Plasmalemma, endoplasmic reticulum and nuclear membrane
Volume of the membrane material (in liquid crystal state}
Vary in the range from 7.5 to 9 %.

Target 3 **GENETIC APPARATUS OF MITOCHONDRIA**

The sizes of mitochondria vary in the range from 0.25 to 5 micron^{3} per one cell.
The number of mitochondria vary from 150 to 500,000 per cell

Figure 6. The main targets of cell injuries bythe action of irradiation.

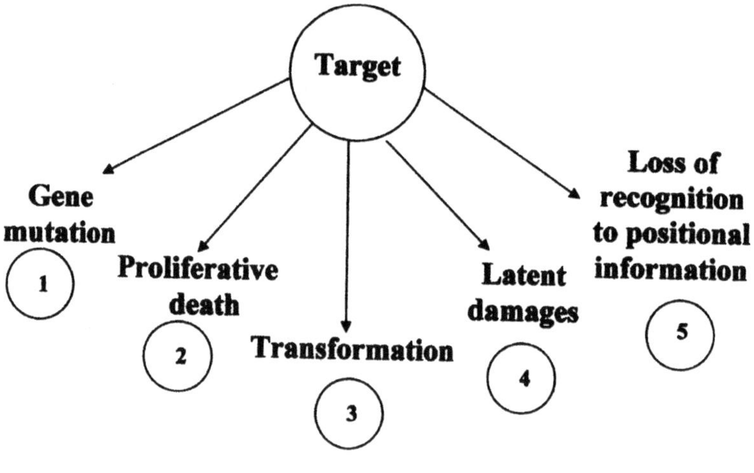

Figure 7. The initial fate of irradiated cells.

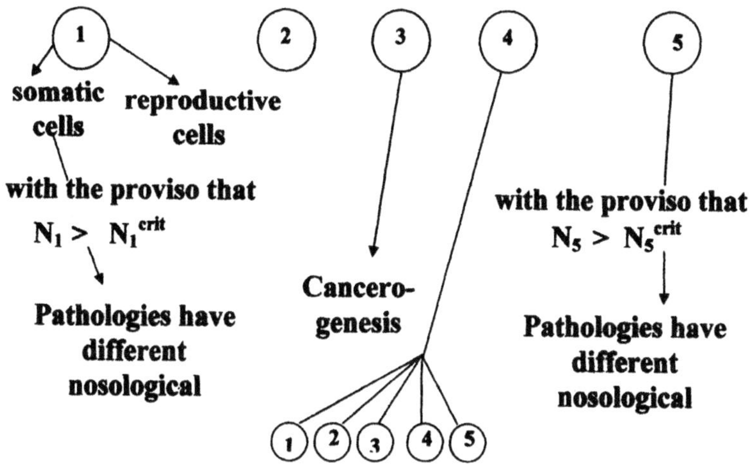

Figure 8. The secondary fate of irradiated cells.

Brief mention should be made of the methods aimed at the mitigation of the consequences of a nuclear explosion. These methods are many and varied. They are named as countermeasures.

The challenge of the countermeasures is designing a reduction of the dose load on people and minimization of ecological harm related to economic function.

All countermeasures should be evaluated by a cost-benefit analysis. This analysis should take into account the costs of operations, expected dose scale of the dose adverting, socio-psychological reaction as well as the time factor.

At least five categories of the countermeasures may be classified related to the domains of application:

- Countermeasures in decontamination of sufferers defeated by irradiation and radionuclide contamination.
- Countermeasures in natural and semi-natural ecosystems.
- Countermeasures in agricultural production.
- Countermeasures in hydrological ecosystems
- Technological countermeasures.

Superficial decontamination of a person is no simple task especially when contamination with radionuclides is combined with burns, scalds and open wounds. Evacuation of people immediately after an explosion from the contaminated area is the most important countermeasure. The decision about evacuation should be made on the basis of operative radiation monitoring.

Reduction of aerosol activity of air by prevention of dust formation has been considered as one of the main tasks at mitigation of the release of radionuclide. The chemical method of dust suppression is based on the treatment of dust forming objects (earth roads, earth shoulders of roads etc.) with the chemical agents that can ether form a crust of the treated surface or accumulate the dust. Latex, a composition of hydrolyzed polyacrylonitrile, lignosulphonates and oil residue were used after the Chernobyl Accident. The application of these materials serves as countermeasures of short duration. A more complex problem is the long-term chemical or biological fixation of dusty areas. The cultivation of some perennial plants had beneficial effects in conservation of the dust material. The root systems and above ground parts of plants protect the surface of sandy soils from deflation and the secondary transportation of radionuclides.

Implementation of countermeasures in agriculture on contaminated territories is a very important action related to radiation protection.

Countermeasures in crop production are mainly based on changes in land use. The most effective countermeasures are as follows: decreasing the area of crops characterized by high accumulation of radionuclides, strictly abandoning the area of very high contamination, and removal and transportation of the upper 5 cm thick layer of soil.

The some technologies of phytodecontamination or phytorehabilitation were tested. It was found that the usual practice of cultivation of plants and gathering of phytomass had little to offer as a method of soil deactivation as to ^{137}Cs and ^{90}Sr because these radionuclides are provided by great masses of non-isotopic carriers in soils. However, higher efficiency technologies were developed, namely, rhizofiltration and combining of growing plants characterized by high capability for dissolving of radioactive particles in soil and washing the plough layer of soil with succeeding cultivation of some water plants.

Agrochemical countermeasures consist of liming acid soils, application of increased doses of mineral fertilizers, supplementiong soils with natural absorbents and using of organic fertilizers.

Inasmuch as the main source of inner irradiation is radionuclides that are contained in milk and meat, it is evident that countermeasures in stockbreeding play a leading role in decreasing the dose load on people. The restriction on the use of milk foodstuffs produced in territories of high levels of radionuclide contamination is the most effective countermeasure. The application of some compounds preventing the absorption of radionuclides in the gastro-intestinal tract when these substances are introduced into the diet of animals is also a very efficient countermeasure.

Decontamination of the Hydrological Ecosystems are carried out by phytoremendiation or rhizofiltration. Some water plants, algae as well as the higher plants, are exemplified by the in capacity for absorption of radionuclides in great quantities.

The efficiency of the countermeasures is dependent on a number of factor, namely, radionuclides composition, post-explosive time, environmental conditions (type of soil and landscape), ecological acceptability and social perception. The latter, incidentally, is a very important factor.

The main countermeasures are classified in *Table 6.*

Table 6. Mitigation of the consequences of the nuclear terrorism.

DECONTAMINATION OF THE ENVIRONMENT	MEDICAL PROTECTION	SOCIAL PROTECTION
Decontamination of territo-ries(removal, sorbtion, phyto-rehabilitation and others) Phytoremediation of hydrological system Water quality control Deep processing of conta-minated foodstuff and wood Altering the type of lands using etc.	Decontamination Evacuation Assessment of dose Control of drinking water and foodstuff water and foodstuff Health monitoring Decreasing of the risks of late effects Retrieval of immune system Therapy of radiation disease	Psychological prophylaxis Adoption of worthy laws and regulations System of compensation Health monitoring Explanation the real situation etc.

During sixteen years serious efforts were mounted to develop - reliable countermeasures as applied to various situations in the territories around the Chernobyl Nuclear Power Station.

LITERATURE

1. *Chornobyl Catastrophe*/ Ed. By V.G. Baryakhtar. - Kyiv: Editoral House of Annual Issue "Export of Ukraine", 1977. -576 p.
2. *Research Activities about the Radiological Consequences of the Chernobyl NPS Accident and Social Activities to Assist the Sufferers by the Accident* / Ed. By Imanaka. -Research Reactor Institute, Kyoto University, 1998. -278 p.
3. Conference "*Health Effects of the Chernobyl Accident: Results 0f 15-Year Follow-up Studies*". *June* 4 - 8, 2001. Kiev, Ukraine. Abstracts. In: International Journal of Radiation Medicine. Special Issue, 2001, Vol.3, No 1 - 2.
4. *The National Atlas of* Ukraine (electronic version) Ed.by L.G.Rudenko, et al., Kiev, 2002

Environmental Security: A Case of Slovenia

Marjan Malesic
Defence Research Centre, Faculty of Social Sciences, University of Ljubljana, Slovenia

INTRODUCTION

Theory on the causal connections between environment and national security suggests that there are three key arguments to be taken into account to explain this relationship: existential, physical and political arguments[1]. However, it seems that the explanatory value of these arguments depends predominantly on the definition of national security and partly on the definition of environment. These arguments and definitions are applied in the analysis to prove that civil society actors in Slovenia perceive environmental threats as crucial for their (in) security, whereas the political and military institutions place the environmental issues into the national security considerations, at least at the declarative level. Nevertheless, the list of environmental threats stemming out of the civilian and military surroundings, and the way of dealing with them, reveals many deficiencies and pitfalls potentially dangerous for the environment, and consequently for the security of the citizens of Slovenia.

THEORETICAL CONSIDERATIONS

As mentioned above, there are three key arguments or forms of connections between the environment and security, which could be (according to Levy, 1995) termed existential, physical and political. **Existential argument** claims that certain aspects of global environment are so closely connected to national values that they represent a security interest. The degradation of environment[2] jeopardizes the most basic aspect of man's security; it brings about the undermining of natural support of systems that are crucial for man's activity and existence. **Physical argument** more selectively holds that degradation of

---

[1] See Marc A. Levy, Is the Environment a National Security Issue? International Security, Vol. 20, No. 2 (Fall 1995), pp. 35-62.

[2] The degradation of environment is seen as accelerated use of natural resources, pollution of the atmosphere (climate changes caused by green house effect, ozone depletion, acid precipitations), depletion of fresh water resources, chemization of environment, degradation of soil and the loss of biological diversity.

K. Mahutova et al. (eds.), Defense and the Environment: Effective Scientific Communication, 139–152.

environment becomes a direct physical threat to security when it causes the loss of well-being and lives of citizens or it jeopardizes basic social values in any other way. Not all environmental threats are of such a nature and/or scope. **Political argument** raises a question, whether the degradation of environment could cause civil riots, mass migrations and consequently armed conflicts in the areas where the absorption capacity of the environment to provide survival to people is rather limited?[3]

It seems that in practice there is a vivid connection between the environment and national security - the main problem in explaining this relationship seems to be of academic nature. Namely, it is a matter of definition of both, the national security and to a lesser extent the environment. The typology of national security definitions in last decades shows that there are two main approaches to be explored: **traditional (deterministic) one and post-modern (complex) one**. The **traditional approach of national security** understands it as the absence of external military threat and as a use of military means to protect from outside (Berkovitz and Bock, 1965: X; Steinke and Vale, 1983: 8, 9; Levy, 1995: 5) and inside (Cohen and Tuttle, 1972: 1) military attack. This group based the concept of national security upon organized violence between or within the states and not beyond that (see Deudney, 1993: 350). The military determinist approach to define national security prevailed during the Cold War and it seems that the omni-presence of the fear related to possible armed (even nuclear) conflict between two major political-military blocks prevented scientists to perceive other important threats.

However, as early as in late sixties, more **complex definitions** of national security appeared. McNamara (1968, 150) warned that national security is not a semantic game, in which we equate security exclusively with military phenomena, even more, with military hardware even if in practice it is not so. Taylor (1976: 3, 4) noticed that the majority of Americans saw national security as a military defence against a military adversary and as a matter armed forces were accountable for. According to him, it was necessary to recognise other threats to security, whereas security should protect all important social values, material and non-material ones. An adequate national security policy should protect social values from military and non-military dangers, external or internal ones. Brown (1982: 21, 22) also warned that apart from traditional military threats, appeared new, environmental and economic threats, for which traditional organization of national defence was useless. Ullman (1983: 133) suggested that a threat to national security is an action or a sequence of events that threatens drastically, and over a relatively brief period of time, to degrade the quality of life for the inhabitants of a state, or threatens significantly to narrow the range of policy choices available to a

---

[3] As to the armed conflict, the role and position of environment could be observed at the levels of its causes (do environment related problems provoke armed conflicts?), its course (are environmental objects used as military targets and/or means?), and finally, its consequences (to what extent armed conflict harms the environment?).

state or to private, non-governmental entities (persons, groups and corporations) within the state. Violence is not only a matter of intense activity, but could be exercised latently on the structural level. Mathews (1989: 162) emphasised environmental and demographic threats, and the depletion of natural resources as a threat to national security. Loske (1991: 25) provided an important input to this strand of thought by emphasising that the threats to security do not emanate only from the relationship between the states but also from the relationship between man and nature. Equally important is his observation that environmental problems pose a threat to security that acts cross-nationally; it goes beyond the sovereignty of national states and requires intense and thorough international cooperation.

An important impact to security thought after the Cold War provided Beck (1992) with his concept of 'risk society'. According to him, the modern society transformed significantly from a society of wealth inequalities in which (although these inequalities remain important) the main problem are environmental dangers that cross traditional inequalities. Typical risks to contemporary society are nuclear and chemical pollutions. Shaw (1996: 28) added that in 'risk society' not only traditional social divisions are crossed, but also political boundaries.

Huge impact on the 'state of the art' brought about a so called 'Copenhagen school of security' in early nineties, which **'reconciled' traditional and post-modern definitions of national security**. The representatives of this school (Buzan, Kelstrup, Lemaitre, Tromer and Weaver, 1990; Buzan, 1991; Weaver, Buzan, Kelstrup and Lemaitre, 1993; Buzan, Weaver and de Wilde, 1998) understood national security as a combination of military, environmental, economic, political and societal (identity and cultural) dimensions. They rejected the traditional concept of security as exclusively or predominantly a military matter, but at the same time included the military dimension in their non-traditional concept of security. Wiberg (1996: 23) endorsed this approach in his considerations of national security.

Grizold (1999: 28) defines security as a state, in which balanced physical, spiritual, psychic and material existence of an individual and social community in relation to other individuals, social communities and nature is achieved. National security encompasses security of state territory, people and their property, maintenance of national sovereignty and realization of basic social functions as social, economic, socio-political, cultural, ecological and other functions (Grizold, 1999: 3).

Nonetheless, Politi (1997: 13) suggested that multi-dimensional approach to national security is not something new. According to him, this approach is typical for Europe during the Cold War, but even at that time, the 'absence' of non-military dimensions in security debate was more a consequence of strict division of external military threats (a domain of the military) and internal threats (a domain of police), rather than a problem of non-existence or non-acknowledgement of non-military threats. He fears the 'inflation of national security concept' could occur if all disintegration processes and dangers become a matter of national security.

Along with new definitions, a question **who should be a primary referent of security** was raised. National security was reconstructed due to the fact that 'nation' and 'state' did not coincide in many cases. Copenhagen school (1993) introduced a term **'societal security'** to emphasize that the primary referent of security was **society**, understood as national, ethnic or/and religious community. Booth (1990) was a representative of radical liberal approach, suggesting that **the individual** should be a primary referent of security. **State** is not understood as an entity that could always provide security, on the contrary, every state should be critically examined to see whether it poses a threat to an individual, provide the security to him or a combination of both. Shaw (1996) had suggested that global environmental challenges requested 'concert' decision-making and common political institutions in international community.

ENVIRONMENTAL SECURITY

Environmental dimension of security is rather young compared to other dimensions. In early seventies the 'Rome Club' warned about the limits of human development and growth. In 1972, the first UN international environmental conference was held in Stockholm and warned that the humankind has reached a historical point at which **we should be careful about the consequences of our activity**. UN Ecological Programme was established soon after the conference and several other international environmental events occurred in a decades to follow, among them international conferences on environment in Rio de Janeiro (1992) and Kyoto (1997). Several international conventions, declarations and agreements have been stipulated and signed since, whereas the positive effects do not seem to be (entirely) satisfactory.

Simultaneously with the considerations on the threats to national security stemming out of the relationship between man and nature, the term **'environmental security'**[4] entered the security-related debate. Brown (1977) connected environmental security with fields like energy, biological systems, climate changes, nutrition dangers and economic threats. Buzan (1991: 19) saw environmental security through maintenance of local and planet biosphere as basic systems upon which all man's activities are based. Stockholm International Peace Research Institute (SIPRI, 1992: 27) defined environmental security as a situation in

[4] Terriff, Croft, James and Morgan (1999: 182) suggested that there is a difference between 'environmental security' and 'ecological security': ecological security refers to the protection of environment without taking into account man's needs, whereas environmental security refers to the relationship between the protection of environment and man's needs.

which governments alone or with international help are capable of mitigating social and political impacts of natural resources depletion. Buzan, Weaver and de Wilde (1998: 84) saw environmental security as a status of ecosystems, which are the basis for maintenance of achieved level of civilization. Chalecki (2001: 2) held that the environmental security is a capability of nation and society to cope with environmental risks and harmful changes, and controversies or conflicts related to the environment. On the other hand Terriff, Croft, James and Morgan (1999: 132) criticised conceptualisation of environmental security by suggesting that it is not appropriate to militarise environmental problems.

PUBLIC AND OFFICIAL PERCEPTIONS OF THREATS IN SLOVENIA

In the following pages, I want to present the public and official perceptions of threats in Slovenia by emphasising the role and meaning of the environmental dimension of these social, psychological and political processes. It is my intention to demonstrate that **the environmental threats and environmental security in general play an important role in security related considerations** that have been taking place in our community in a last decade or so.

Public Perception of Threats

In theory, we speak about 'real threats' – military and political situation in the surroundings of a certain community, frequency and power of natural and technical disasters, environmental conditions etc. – and about 'perceived threats' that are subjectively perceived by the individuals and are reflected in public opinion. It means that national security system functions in two security 'surroundings', in 'real' and 'perceived' surroundings: the first one is the realm of practice, in which the national security system endeavours to provide effective security, defence and protection of community against different threats that jeopardize its cultural, natural and material values and interests; the second one is the realm of perception, of the attitude toward national security system and feelings about (in)security related to its existence. The national security system thus functions in the interaction with the 'double' surroundings: first, the surrounding of internal and external threats to security, the original reason for the existence, structure and organisation of national security system, and second, the surrounding of the security as a value, in which the image of national security system influences popular perception of security and their trust in different parts of the system. It means, security is not only a 'matter of strategic affairs' but also a 'matter of spirit' – states, social groups and individuals need real security and the feeling of security.

If we present the above deliberation in a more operational way, we could say that the level of real and perceived threats are mostly dependent on the relationship between the dangers brought about by different threats and readiness of community to deal with these threats in an organized, systemic and effective manner.

As seen in Table 1, respondents perceived **non-military threats** to be crucial for the security of Slovenia: drugs and narcotics (the negative phenomena of the 'westernization' process are more drug users in our society), crime (the crime rate is getting higher every year, especially the organized international crime dealing with drugs and illegal immigrants), traffic accidents (around 300 people die on the roads every year in Slovenia), unemployment (it is decreasing in recent years, it reached app. 7 per cent in 2002 according to International Labour Organisation-ILO standards), **degradation of environment** (nuclear power plant, chemical industry, industrial waste, global environmental problems) and poverty (generally speaking Slovenia is a relatively rich country – comparable to Greece and Portugal, members of EU – but the society is under a process of more emphasised stratification, which jeopardizes some social strata in socio-economic terms).

We can also see from the table that terrorism and military threats are not perceived as important ones, which is a confirmation of a trend experienced in the last decade in Slovenia. It is also worth mentioning that the average figures of threats in 2001 are lower than in 1999, meaning the society felt safer in general terms. On the other hand, the role of **degradation of environment** as a threat is relatively less prominent in last couple of years than five and ten years ago, respectively. To be more precise, the average values are comparable throughout the entire period (slightly above mean value 3), whereas the place of the degradation of environment among other threats is lower in last years than before (1990 – second place, 1994 – first place, 1999 – fourth place, 2001 – fifth place).

It is necessary to stress though, that some **military and non-military threats are intertwined**, which brings us to the concepts of 'complex threats' or 'inter-woven threats'. It is plausible that the degradation of environment could cause poverty, mass migration and the latter could cause frictions and armed violence. On the other hand, armed conflict always brings other security problems: mass migration, deterioration of socio-economic situation, drug and arms trafficking, higher probability of terrorism, degradation of environment, and general deterioration of living conditions.

Table 1: Perception of threats (Slovene Public Opinion 1990/91, 1994/95, 1999, 2001)

	1990	1994	1999	2001
drugs, narcotics	-	2.92	3.46	3,41
crime	2.75	3.11	3.46	3.28
traffic accidents	-	-	3.22	3.24
unemployment	-	-	3.35	3.14
degradation of environment	***(2)3.36**	**(1)3.17**	**(4)3.35**	**(5)3.07**
poverty	-	-	3.13	3.05
low birth rate	-	2.26	3.39	3
economic problems	3.46	3.08	3.23	2.99
suicides	-	-	3.09	2.88
sell-out of social property	2.91	3.02	3.14	2.87
natural and technological disasters	**3.1**	**2.81**	**3.2**	**2.76**
refugees, illegal immigrants	-	2.69	2.98	2.74
internal political instability	-	2.89	2.94	2.53
contagious diseases, AIDS etc.	-	-	2.78	2.43
lagging behind in the field of science & technology	3.19	2.67	2.84	2.33
extreme nationalism	-	2.49	2.54	2.2
terrorism	-	2.46	2.64	2.09
conflicts on the territory of former Yugoslavia	2.94	2.73	2.75	2.09
military threats of other countries	1.65	2.37	2.21	1.79

Source: The archives of Defence Research Centre, University of Ljubljana

Explanations: a mean value on the scale: 1–not a threat at all, 2–weak threat, 3–medium threat, 4-strong threat

*Place of threat in the scale of a certain year and the mean value.

Official Perception of Environmental Threats

The Constitution of Republic of Slovenia gives a right to its citizens to enjoy **a healthy living environment** and it is a duty of the state to take care about it[5], meaning, the indirect relationship between environment and security has been established in Constitution. In 1995, the Environment Protection Act was adopted by the Parliament, providing the purpose, system, objectives and principles of the environment protection. On that basis, the National Environment Protection Programme was adopted in 1999. The same Act was also a basis for the Environment Protection

[5] See Constitution of the Republic of Slovenia, 1991, p. 24, Article 72.

Programme in Slovene Armed Forces in 1998. This is the constitutional and legal basis, whereas in the following lines, I want to present the status of **environmental threats in the strategic and doctrinal national security documents**.

The Resolution on Strategy of National Security of Republic of Slovenia adopted by the Parliament in 2001 holds that the importance of military threats is diminishing, whereas non-military threats, risks and challenges, which are equally dangerous to contemporary societies and states as war, gain a momentum. These threats, risks and challenges are often inter-related and interdependent, and they act transnationally on political, military, economic, information, **environmental** and internal security domains.

More specifically, threats posed to the society and state could be subversive activity, threat with aggression, military attack, mass migrations, terrorism, organized crime, **degradation of environment**, economic blockades and **energy crisis**, information and cyber blockades, medical and epidemic threats, and **natural and man-made disasters. Environmental risks for the state are chemical, radioactive and other sorts of pollution, and uncontrolled interventions of man into natural processes**. The negative effects could be seen in global climate changes, damaging genetic consequences, animal and vegetal diseases, unhealthy food and water, and cross-national influences of environmental disasters. The constant threat to national security of the state and its citizens is a possibility of **huge disasters that have their origins in natural processes or in man's activity**.

The Defence Strategy of Republic of Slovenia's chapter on "Security Surroundings-Security Challenges" addresses the environmental security by saying that civilization challenges are **over-burdened environment and consequently threats to living environment**, organized crime, illegal migrations, and ethnic and religious controversies in a broader security surroundings.

Civil Defence Doctrine's chapter "Changes in security surroundings-change of threats after the Cold War" claims that after the Cold War, we deal with other dangers that were not perceived as threats in the past, however, they could, similar to war, also jeopardize contemporary state. These threats are **constant global growth of population, the effects of industrialisation on biosphere, uncertainty about the future procurement with energy**, vulnerability of information and communication systems, **problems of agricultural production and procurement**, instable international financial markets, economic conflicts and similar threats.

The Civil Protection Doctrine's chapter "Dangers and threats to people, animals, property, cultural heritage and environment" holds that among those dangers that pose **a threat to environment are unreasonable exploitation of natural resources, un-thoughtful interventions of man into environment and different forms of**

pollution that ruin natural balance and jeopardize healthy life of man and the existence of animal and vegetal species. Very dangerous are interventions into the environment, which change the landscape, build urban and industrial systems, changes of water regime, introduction of mono-cultures in agriculture and forestry, and the exaggerated use of chemical substances for protection of plants, fertilizers and alike.

The analyses of above mentioned documents revealed that the phenomena of environment and security are closely inter-related and inter-dependent in the official considerations about national security. The environmental security in its broadest term became even a constitutional matter in Slovenia. Evidently, degradation of environment poses a threat to national security, and to this effect, all dangers, risks, challenges and pitfalls stemming out of environment are enumerated and taken into account in the key national security documents.

KEY ENVIRONMENTAL PROBLEMS IN CIVILIAN AND MILITARY SURROUNDINGS

Have the above mentioned popular and official perceptions had any positive impact on **the protection of the environment** in Slovenia so far? According to empirical environmental data, in the mid-nineties, Slovenia was placed in a group of moderately polluted European countries. However, the degradation of environment was higher than expected on the basis of the achieved level of economic development and GNP per capita. The **main problems** seemed to be and still are (Plut, 1998):

- pollution of air with sulphur dioxide,
- acid rain as a consequence of too much sulphur in the atmosphere,
- natural resources are more or less exploited,
- valley ecosystems where majority of people live are extremely burdened,
- biotic and landscape varieties are diminishing,
- nuclear waste problem from the nuclear objects[6] (nuclear power plant, nuclear research reactor and closed uranium mine); the low- and mid-level nuclear waste from nuclear power plant is after almost 30 years of its functioning still at its temporary location in the vicinity of the plant,
- deforestation and degradation of forests,

[6] The responsibility to control nuclear objects is primarily vested with the Agency of Republic of Slovenia for Nuclear Safety.

- pollution of water resources despite the reduction of waste waters[7],
- huge quantities of produced waste.

The situation in the last decade have been improving, partly because of higher legal standards and the development of environmental consciousness, and partly because of the fact that many heavy industry enterprises ceased to function due to economic reasons. The only option for Slovenia in European Union seems to be a concept of sustainable development with high environmental standards.[8]

And what about **the armed forces** - do they respect environmental principles in every day life? In 1998, the General Staff of Slovene Armed Forces (SAF) approved a Programme for Protection of Environment. The Programme determined the organization and tasks, goals, principles and strategy, and financing of the protection of environment in SAF. However, our analysis[9] revealed that the Programme was a hot topic of more or less favourable discussions in SAF in 1998, much less in 1999, while in 2000 there was very little discussion on the topic. The military organization is adjusted to the environmental needs. There are deputy commanders responsible for the protection of the environment at the level of operative commands and brigades, while other commands and units have military officials responsible for the matter. On the other hand, soldiers are not taught and trained in protection of environment and there are no 'ecological units' available in SAF. In practice there are some improvements in certain military barracks and training polygons, nevertheless, there are still many deficiencies as far as the protection of environment in SAF is concerned. In **the military barracks** the situation is as follows:

- military activities still produce a lot of noise,
- waste is collected separately and taken care of by the civilian companies,
- parking lots of military vehicles are still a huge danger due to inadequate surface and a danger of hazardous material spill-over in the underground waters,
- cleaning devices are a rare case in military barracks,
- the military barracks in independent and democratic Republic of Slovenia are being ousted from the urban surroundings where certain elements of environment protection are better,

---

[7] In mid nineties, only half of Slovenian population (total population of Slovenia is a bit less than 2 million) used sewage system and only 18 per cent of population used cleaning devices.

[8] Slovenia will enter European Union in May 2004.

[9] The analysis was accomplished by a group of post-graduate students of Defence Studies at the University of Ljubljana. The author was a coordinator and supervisor of the group.

- the military are more and more dependent on the civilian services (e.g. medical service except the first aid), therefore they produce less hazardous waste.

The situation at **the training polygons** is even worse:

- military activities produce a lot of noise,
- air pollution in the area is quite frequent and intensive,
- polygons are not kept properly, there are no proper parking lots and toilets available (sometimes soldiers bring chemical toilets with them), there is no sewage system, no electricity, while the water is provided in tankers,
- fires occur frequently,
- waste at the polygons is piling up (artificial targets, tyres, redundant wood etc.),
- simulation technique is not used to the extent it could be to reduce real training with degrading effects to the environment.

We also noticed that the residual military culture typical for communist times according to which, everything is allowed for the defence of homeland and consequently there are no limitations for the armed forces, is to a certain degree still there. Ecological conscientiousness has not been developed yet in the SAF to the satisfactory level. Fortunately, the size of Slovene armed forces, and consequently their activities are rather modest; therefore, the country does not face problems related to nuclear, chemical and biological weapons, nor problems emanating from extensive use of conventional weapons in the military training process.

CONCLUSION

We could say that perceived threats of Slovenian civil society (especially the public) and political (official) elite are in conformity with the reality. After the end of the Cold War and the end of armed conflicts in the former Yugoslavia, military threats retreated and non-military threats took a lead. SIPRI[10] data indicated that the number of armed conflicts in the world is decreasing, while among them, those of internal character prevail (intra-state conflicts). Non-military threats become more important for the security of the contemporary states, although they existed before, but it seems that only lower plausibility of the huge military conflict enabled us to perceive their importance for the security and stability of national states and international community. These threats refer to economic and social field, environmental issues, mass migration,

[10] See SIPRI yearbooks from 1996 on.

health and epidemiological conditions, terrorism, illegal drug and arms trafficking, and also natural and technical disasters.

The analysis revealed that at the declarative level the environmental security gained a lot of momentum in the last decade in Slovenia, and we could say that **existential argument** on the relationship between security and environment is more than evident in public attitudes and official documents. However, in practice there are still many environmental problems and a brief analysis of civilian and military surroundings proved the introductory assumption.

It seems that environmental issues in Slovenia still have a useful mobilization effect for the political elite and civil society, whereas in reality the practical measures to improve the environmental situation do not necessarily follow this pattern of mind.

REFERENCES

Plut, Dusan, *Varstvo geografskega okolja.* Ljubljana: University of Ljubljana, 1998.

Plut, Dusan, *Brez izhoda? Svetovni okoljski procesi.* Ljubljana: DZS, 1995.

Ministry of Defence, General Staff of Slovene Army, Environment Protection Programme of Slovene Army. Ljubljana, 1998.

Marc A. Levy, Is the Environment a National Security Issue? International Security, 1995; 2: 35-62.

Berkowitz, Morton, Bock, P. G., *American National Security.* New York: Free Press, 1965.

Steinke, Rudolf, Vale Michel, eds. *Germany Debates Defense.* Armonk: M. E. Sharp, Inc., 1983.

Goldfarb, Theodore D., ed. Taking sides. Clashing Views on Controversial Environmental Issues (5th edition). Guilford: The Dushkin Publishing Group, Inc., 1993.

Cohen, Ira S., Tuttle, Andrew, S. *National Security Affairs: A Syllabus.* New York.

McNamara, Robert, The Essence of Security. New York: Harper & Row, 1968.

Brown, Lester, R., "An Untraditional View of National Security". *In American Defense Policy*, Reichert, J. F., Sturm, S. F., eds. Baltimore: The John Hopkins University Press.

Loske, Reinhard. Ekoloski kraj nacionalnih drzava. Medzunarodna politika 1991; 982: 24-9.

Taylor, Maxwell, D. *Precarious Security.* New York: Norton, 1976.

Ullman, Richard. Redefining Security. International Security, Vol 8. No. 1.

Mathews, Jessica T. Redefining Security. Foreign Affairs, Vol. 68, Spring.

Shaw, Martin. "Nova varnostna razprava in družbene spremembe". In *Razvoj obramboslovne misli.* Marjan Malesic, ed. Ljubljana: University of Ljubljana, 1996.

Booth, Ken. Security and Emancipation? Review of International Studies, 1991; Vol. 17, No. 4.

Buzan, Barry, Kelstrup, Morten, Lemaitre, Pierre, Tromer, Elizabeth, Waever, Ole, *The European Security Order Recast: Scenarios for the Post-Cold War Era.* London: Pinter, 1990.

Buzan, Barry, *People States and Fear: An Agenda for Post-Cold War Era.* London Harvester Wheatsheaf, 1991.

Buzan, Barry, Waever, Ole, de Wilde, Jaap, *Security: a New Framework for Analysis.* London: Lynne Rienner, 1998.

Waever, Ole, Buzan, Barry, Kelstrup, Morten, Lemaitre Pierre, *Identity, Migration and the New Security Agenda in Europe.* New York: St. Martin Press, 1993.

Chalecki, Elisabeth L., Environmental security: a case study of climate change, http://www.pacinst.org/environment_and_security/env_security_and_climate_change.pdf; 2001.

Wiberg, Hakan. "Security Problems of Small Nations". In *Small States Security in the New Europe,* Werner Bauwens, Armand Clesse, Olav Knudsen, eds. London, Brassey's, 1996.

Grizold, Anton, *Evropska varnost.* Ljubljana: University of Ljubljana, 1999.

Politi, Alessandro, European Security: The New Transnational Risks. Chaillot Papers, No 29.

Brown, Lester. Redefining Security. Worldwatch Paper No. 14. Washington, 1977.

SIPRI Yearbook 1992, Stockholm.

Terriff, Terry, Croft, Stuart, James, Lucy, Morgan, Patrick, *Security Studies Today.* Cambridge: Polity Press, 1999.

The archives of Defence Research Centre, University of Ljubljana.

Constitution of Republic of Slovenia.

National Environment Protection Programme.

Environment Protection Programme of Slovene Army.

Resolution on the Strategy of National Security of Republic of Slovenia.

Defence Strategy of Republic of Republic of Slovenia.

Civil Defence Doctrine of Republic of Slovenia.

Civil Protection Doctrine of Republic of Slovenia.

Environment Protection Act of Republic of Slovenia.

Connelly, James, Smith, Graham, *Politics and the Environment: From Theory to Practice*. London: Routledge, 1999.

Ferry, Luc. *Novi ekoloski red. Drevo, zival in clovek*. Ljubljana: Krt, 1998.

Pepper, David, *Modern Environmentalism. An Introduction*. London: Routledge, 1996.

Garner, Robert, *Environmental Politics*. Englewood Cliffs: Prentice Hall, 1996.

Mazur, Allan. Global Environmental Change in the News. International Sociology 1998; 4: 457:72.

Beck, Ulrich , *Risk society*. London: Sage, 1992.

Doyle, Timothy, McEachern, Doug, *Environment and Politics*. London: Routledge, 1998.

Military Conversion/Restructuring and Sustainable Development Strategy

William Zadorsky
Ukrainian Ecological Academy of Sciences, Ukrainian State University of Chemical Engineering., Dnepropetrovsk, Ukraine

INTRODUCTION

The Ukraine is experiencing an extremely difficult ecological situation standing on the brink of environmental collapse. Further aggravation of the current situation is likely and can be explained by the transition to the mechanisms of a market economy and management when destabilization of an old system and industrial disruption result in the accumulation of vast quantities of toxic wastes. The policy of property privatization, transition of the defense military doctrine in the Ukraine, as well as a reduction in the military output result in the deterioration of valuable material-technical, intellectual and human resources, and decline in industrial capacities.

DISCUSSION

Taking into account the above, one of the priority directions for defense industry conversion is the provision for reliable development of the state on the basis of a Ukrainian environment strategy. This strategy should be directed at industrial and agricultural development along with the implementation of the concept for adaptation and rehabilitation of the population and introduction of effective systems for life support under adverse environmental conditions. Priority direction for conversion of military enterprises is demilitarization of territories and military objects and provision of sustainable development on the basis of: "greening" of industry, adaptation and rehabilitation of the population, and implementation of effective life support systems in the conditions of an adverse environment.

K. Mahutova et al. (eds.), Defense and the Environment: Effective Scientific Communication, 153–158.

It is necessary for the Ukraine:

- to accelerate elaboration of the Program for sustainable development in accordance with the Rio de Janeiro summit decisions of 1992 signed by the Ukraine
- to elaborate and introduce the Environmental Conversion Program as a special part of the Program for Sustainable Development of the Ukraine;
- to undertake structural reconstruction and conversion the military industry in regions with unfavorable environmental situations on the basis of sustainable development;
- to set priorities in the state policy of sustainable development and military conversion for the design and manufacturing of products related to the environmental protection and national use of natural resources, environmental monitoring and survival of the population in critical technogenic conditions;
- to accelerate adoption of the State Program for Industrial and Domestic Wastes Utilization with the consideration of the military complex potential and capacities;
- to consider the concept of industrial symbiosis and inter-industry cooperation as fundamental in defining ways for industrial and domestic waste processing and recycling;
- to render as expedient the development of local and territorial monitoring systems for military industrial objects which present a threat to environment and health of the population and for the objects of the nuclear fuel cycle;
- to consider environmental security as an integral part of the National Security of the Ukraine in regard to pressing environmental programs and international obligations taken by the Ukraine;
- to initiate appearance and development of the Ukrainian Scientific-Technical Environmental Conversion Program using the experience of European countries in the development of environmental conversion programs and optimal methods for management and financing;
- to consider the importance of economic substantiation in developing and implementing conversion capacities for solving environmental problems; and
- to pay particular attention to economic aspects of military-related environmental problems.

It is very important to approve activities of the NATO Program on Priority Area on Environmental security in consolidating efforts of NATO country scientists and specialists focused on rendering assistance in the solving of global and regional environmental problems of sustainable development. The second half of twentieth century was characterized by unprecedented progress in science, industry and economics in some countries accompanied with not less unprecedented growth of population, mainly in the other countries. The price that mankind has paid is progressive destruction of natural resources and increasing contamination of the environment. Economic and environmental problems are tightly intertwined with the social ones. The development of independent countries, formed at the break-up of Soviet Empire, has in addition some specific peculiarities connected with radical changes of their social and economic systems. Such changes are for the FSU countries rather painful and lingering. So, the Ukraine has not overcome in over decade its "transitional state" in its economy, as well as in social and political life.

Up to now the country remains in a state of vagueness with respect to a lot of fundamental questions. What is our future? How long shall we live in a country with the so-called "transitional economy"? What type of social and economic system will be formed finally in our country? How will we solve our economic, social and environmental problems? How can we ensure sustainable balance between human beings, industry and the environment? The Ukraine still has not adopted the concept of sustainable development, which today has become a basis for the solution of these problems in many other countries.

Nowadays the idea of sustainable development (SD) of mankind, formulated in 1992 by the International UN Summit in Rio de Janeiro seems to be attractive to many people worldwide. However, the very concept of SD is being taken differently in various societies. Developed societies accept it readily with a hope to retain their leading positions in the developing world for an indefinitely long time. Less developed societies put their hopes on SD to join the advanced group of societies, to a great extent by borrowing their technological and social experience. And finally, the poorest societies, which have no real chances to raise significantly their living standards in the foreseeable future, take the SD concept rather skeptically.

The concept of SD corresponds to the goals of humanity (good quality of life, conservation of environment and economic prosperity). It is worthwhile to put a lot of effort into it. It remains uncertain, however, whether these aims are attainable only in principle, or whether the passionless and implacable laws of nature will set certain limits to our pious intentions. The laws of thermodynamics set insurmountable limits to the efficiency of heat engines. A thorough theoretical analysis is necessary to comprehend these natural limits. The concept of sustainable development includes the following three basic components: economic, environmental, and social. Only balanced progress in all three directions, the growth of the economy tightly linked with the simultaneous improvement of environmental conditions and the solution of social problems, will promote achieving the goals of sustainable development.

The following conditions are necessary prerequisites determining the possibility of sustainable development of mankind:

- decrease of the rate of growth of the world population;

- gradual replacement of existing technologies by cleaner and less energy and resource consuming ones;
- reduction of usage of non-renewable raw materials in favor of renewable ones;
- promotion of sustainable development at all levels of administrative regulation and management (local, regional, state and global).

Recognizing the concept of SD as inspiring and humane, one should not accept it by faith or as an indisputable dogma, because SD contains a series of inherent contradictions. Disclosing these contradictions, analysis of them, and a search for the ways of overcoming them seem to be crucial for sustaining the very concept of sustainable development.

Up to now most countries with transitional economies have had no consistent sustainable development programs of their own. The time has come to make the transition from fruitless talks about some peculiar paths of development to more systematic approaches to realization of concepts and methods that had shown their efficiency in prosperous and stable countries.

It is possible to agree with the necessity of:

- carrying out systematic theoretical studies directed to the main positions of SD, to determining its natural limits and defining the conditions of stability of the process;
- carrying out systematic studies of the laws of evolution of multidimensional, hierarchical and extremely inhomogeneous systems;
- making for better understanding between natural scientists and social scientists and encouraging their joint efforts in studying processes of evolution of human society and in search for the ways of solving problems that arise;
- elaboration and implementation in a global scale the measures for reducing the rate of population growth,
- elaboration of the mechanisms and means of discussion, coordination, and acceptance of decisions concerning the problems of development in various domains (economics, ecology, and social problems) and at various hierarchical levels (local, city, region, country, geographic region, and global);
- elaboration of the mechanisms and means of resolution of the conflicts inevitably arising in any complex, inhomogeneous system in the course of its development;
- promoting investigations and engineering programs directed towards making technologies environmentally friendlier, cleaner production and reducing of usage of non-renewable raw materials;
- promoting research and engineering programs directed to the reduction of energy consumption;

- elaboration and implementation of complex of measures to make the concepts and aims of SD clear for most of the population throughout the world; and
- initiation of discussion and acceptance by the General Assembly of UN of a total ban on export of hazardous prods and wastes, especially radioactive ones, from the countries where they are produced.

To use SD concepts to military ecology problem solving it is necessary to create a network of service organizations. With an eye on the future of our society, many laws have been passed to protect and clean up the environment. Environmental scientists have to have experience to help with the environmental concerns of any project and provide clients with practical solutions to complex environmental regulations. They have to provide environmental science services to the military organizations, to help clients identify areas of non-compliance and provide solutions that can satisfy the myriad environmental regulations that affect municipalities and businesses. Scientists have to perform endangered/threatened species evaluations to determine the impact a project may have on endangered or threatened species or habitats. The latest technology, as well as time-tested methods, must be used to conduct habitat evaluations and to prepare documentation to meet the requirements of various state and federal agencies. They must work with documentation required to meet agency guidelines, quickly assess environmental issues and identify alternative strategies, thereby enabling clients to avoid costly and time-consuming regulatory obstacles. It is especially important for Environmental Services for the Military to provide environmental regulations and permitting requirements that can be overwhelming to those not familiar with them, to help clients acquire a variety of permits.

It is very important for military service organizations to play a key role in the environment. Their functions provide a variety of benefits to aquatic ecosystems, adjacent terrestrial lands, and to the creatures that depend on them, including man. The matter is that some of the recognized roles include: nutrient removal and transformation; sediment removal; toxicant retention; flood flow alteration; groundwater recharge and discharge; wildlife and aquatic diversity and recreation. The full range of services includes preliminary jurisdictional determinations, boundary mapping, impact evaluations, and regulation training.

Concerning the peculiar problems of the Ukraine and, in part, of other countries with transition economies it is possible to recommend:

- to implement the concept of SD as a rule for action on legislative, administrative and economic levels;
- to promote wide popularization of the basic ideas of SD;
- to carry on interregional environmental and economic policies taking into account both the interests of the regions and those of the state as a whole in their mutual interrelation;

- to promote increasing the outputs of high-added-value products and the efficient utilization of domestic resources;
- to promote on governmental levels the actions directed to conservation of energy and resources;
- to take a set of measures in favor of cleaner production and making a more balanced economy by shifting from production of means of production to consumer goods;
- to encourage organizing and developing of international partnerships for solving actual scientific problems connected with cleaner production, protection of environment, and energy and resource conservation;
- to widen international exchange of information on the basis of contemporary information technologies; and
- to enlarge environmental education and to promote international co-operation.

SUMMARY

The priority direction for conversion of military enterprises is demilitarization of territories and military objects and provision of sustainable development on the basis of: "greening" of industry, adaptation and rehabilitation of the population, and implementation of effective life support systems in adverse environmental conditions. It is necessary for the Ukraine to elaborate and introduce the Environmental Conversion Program as a special part of the Program for Sustainable Development of the Ukraine, to undertake structural reconstruction and conversion of the military industry in regions with unfavorable environmental situations, and to implement the concept of SD as a rule for action on legislative, administrative and economic levels.

PART IV:

ENVIRONMENTAL DECISION SUPPORT AND RESEARCH

Index of Environmental Acceptability of Military Training

Environmental Algorithmic Method for Commanders

Aleš Komár, František Božek, and Jiří Dvořák
Military University of the Ground Forces in Vyškov, The Czech Republic

INTRODUCTION

Military training must consider the quality of the environment with the aim of achieving sustainable training. The index of Environmental Acceptability of Training appears to be a suitable instrument within a commander's environmental management system and for integration into national directives concerning environmental protection.

The objective of the index system developed is to create an integral common standard that would become a universal basis for all training on environmental situations so it becomes a basis for all military branches and various military equipment. By using the method the following can be considered in a systematic way:

- all elements of the biosphere including various ecosystems of the pedosphere, hydrosphere, atmosphere, plants, animals and abiotic factors of nature;
- various types of continuing threats to the environment (air pollution by chemicals and dust, soil and water pollution, mechanical degradation of soil, waste accumulation, both direct and indirect damage to life organisms, impact of noise and vibrations, radiation, landscape deterioration, etc.);
- military activity compliance with national environmental law, law of a host nation and also supranational legislation, eventually with law in other areas (e.g. criminal liability, crisis management).

Commanders remain subject to the law. The carrying capacity of the environment to sustain stress will depend on the size of unit and type of training selected. It is possible to have an individual approach for each training situation and assess any possible adverse environmental impact. Although specific situations can be dealt with according to the condition for the training and quality of the environment, common sample algorithms can be designed.

K. Mahutova et al. (eds.), Defense and the Environment: Effective Scientific Communication, 161–166.

It is possible to design a flexible environmental algorithm for a commander. This algorithm will be drafted in compliance with the extent of acceptable risk according to his assessment. The starting point of the algorithm is the Fire and Explosion Index that is applied in order to define industrial accidents.

APPLICATION OF THE METHOD

The system of analysis is based on the principle of a step-by-step system by which are assessed:

- the source of the risk (material, energy),
- the area of possible impact (impact of the source).

The aim of application in a military context is to assess realistically the environmental risks of training troops in peacetime. Quantitative formulation of risks is based on legislation, so that the activity of a unit is not only acceptable from the environmental viewpoint, but also legally unassailable.

The analysis consists of three levels:

- At first **risks (R)** are analysed and sphere of their possible **impact (I)** is determined;.
- Secondly, **total index values of risk** and **impact (TIVR, TIVI)** are determined and then the **environmental characteristics of individual risks (ECIR)** are designed;
- Thirdly, the acquired results are included in the commander's decision making process.

The procedure is shown in **Diagram No 1**.

Diagram No . Method application stages

Analysis:

Index values of risks:

Implementation:

COMMANDERS ENVIRONMENTAL ALGORITHM (CEA)

1) Analysis deals with both the sphere of **risks (R)**, e.g., activities that are dangerous for the environment by their nature and **impacts (I)**, e.g., environmental components that are exposed to negative impact of the activity and might be affected by them.
2) Index values of risks consist of:
   Basic and additional values. Firstly; basic index value of **risk (BIVR)** and the basic index value of **impact (BIVI)**. Secondly; **additional risk (AR)** or **additional** sensitivity of **impact (AI)** are added to the above mentioned values in further classification. The **total index** value of risk (TIVR), or of impact (TIVI), indicating sensitivity of the environment, is **a sum of basic index value and additional value** (BIVR + AI , or BIVR + AI);
   Result of risk and impact multiplication - **environmental characteristics of individual risk** (ECR), which is a product of the two above mentioned values (TIVR x TIVI). It is also a non-dimentional figure (index). Possibilities of training in the given area are designated by means of the index and a chart of environmental impacts.
3) **Chart of environmental impacts**, characterised by two axes, is used in the process of inclusion into a commander´s decision making process.
   Axis X is an axis of impact, so it includes possible impact on the environment, axis Y is an axis of sources, so it includes risk and risky activities. First line includes total factor of sensitivity, the first column includes total factor of risk of source.
   Completion of the chart of environmental impacts is based on the fact that individual elements of impact displayed on the axis X and elements of risk displayed on the axis Y have their point of intersection in a particular cell. Index values are added in the direction of each axis and the results are written into the assessment column (line) that is marked as "sum" i.e evironmental characteristics of individual risk (ECR). Cells marked by a cross (x) are not completed. An example of a chart of environmental impacts designed for operation materials and their risks and impact on waters is shown in *Chart No 1.*
4) An appropriate explanatory scale can be attached to the values (indexes) that do not have any environmental characteristics of their own. An example of an explanatory scale is shown below:

Value (index) is in interval	Level of limitation
<1-2,5>	**A**
(2,5-4)	**B**
<4-9,9>	**C**

Chart No 1: *Chart of Environmental Impacts - Example of Completion (partial)*

		Group		Waters BIVI **1,0**			Soil BIVI **1,0**	
		Subgroup		Special		Other	Arable	etc.
		Category		I. AI **1,5**	II AI **1,0**	**AI 0,0**	I.AI **0,0**	
Group	Subgroup	Category	**Sum**	2,5	2,0	1,0	1,0	
Operational materials BIVR **1,5**	Fuels	I. AR **0,0**	1,5	**3,75**	x	**1,5**	x	
		II. i.e. **0,33**	1,83	**4,57**	x	**1,83**	x	
	Other operational materials	I. AR **0,0**	1,5	**3,75**	x	**1,5**	x	
		II. AR **0,2**	1,7	**4,25**	x	**1,7**	x	
	Chemical substances	I. AR **2,0**	3,5	x	x	x	x	
		II. AR **1,5**	3,0	**7,5**	x	**3,0**	x	
Waste waters	Municipal	I. AR **0,0**	1,0	**2,5**	x	**1,0**	x	
	etc.							

There is a three-level scale in this example. Training on **level A** can be conducted without any limitation. Even risky activities can take place in the area within common security measures. There is a certain risk resulting from the conduct of training on **level B**, which in practice means that it is necessary to take increased precautions with regard to the type of risk. It is suitable to consult a military environmental advisor about activities and their location in the area and if need be to provide emergency measures in order to minimize the impacts of an emergency situation.

As for **level C**, it is not possible to conduct planned activity in the given locality. It is necessary either to search for other areas or to modify training.

METHOD OF DETERMINING INDEX VALUES

Some level of **risk** arises from the training itself and commanders are usually aware of it. Values of **impact** on the environment marking its sensitivity originate from legal regulations and customs with international, national, regional and local validity.

Total Index Value of Risk (TIVR)

TIVR represents possibility of risk arising as a result of the use of hazardous materials or no environment friendly activities. It consists of basic index value of risk (BIVR) and additional risk (AR). It is possible to define five basic groups of risks that are individually assessed with help of BIVR:

Operational materials	**1,5**
Waste waters	**1,0**
Movement of troops	**1,0**
Wastes	**2,0**
Training of troops	**1,5**

Individual groups are further subdivided and additional risks (AR) are added. Additional risk may represents risk for environment with regard to the character of substances (e.g. toxicity), their use (frequency), quantity, etc. The figures express the values of hazard starting from zero (no risk).

Total Index Value of Impact (TIVI)

TIVI represents the risk of impacts on the environment with either general or special protection of their components. It seems to be advantageous to start from the laws regarding environmental components when determining these components and their sensitivity. Total index value of impact (TIVI) consists of basic index value of impact (BIVI) and additional sensitivity of impact (AI).

It is possible to define e.g. eight basic groups that can be exposed to possible impact. They are assessed by BIVI on e.g. waters, soil, forests, fauna and flora, geologically and archeologically significant places, specially protected areas, cultural relics and residential areas. BIVI may equal **1,0** in all cases.

Individual groups are further subdivided and additional sensitivity of impact (AI) is added. Various subgrups are determinated mainly according to significance of environmental protection, the importance of biotope, etc. (zero - impact on common natural sources).

REFERENCES

1. Dow's Fire and Explosion Index Hazard Classification Guide, AICHE New York, 1994, ISBN 0816906238.
2. DAVÍDEK, I. Graduation Thesis, Technical University, Ostrava, 2000.
3. KOMÁR, A. at all. Index of Environmental Acceptability of Training. Record of Proceedings. NTG/ASG ETWG. Aachen, Germany. September 2000. Anex O.
4. ŘEHÁK, D. and KOMÁR, A. Index of Environmental Acceptability of Military Activities. Proceedings of the Economy, Logistics and Environment in the Armed Forces III. 2003 April 29 - 30; Brno, Military Academy.

Method for the Military Facilities Risk Assessment

František Božek, Aleš Komár, and Jiři Dvořák
Military University of the Ground Forces, Faculty of State Defence Economy and Logistics Vyskov, Czech Republic

INTRODUCTION

Risk management is a decision-making process issuing from the results of risk analysis. Risk is a term expressing the relationship between the weight of event's consequences and the probability of its occurrence. Individual risk is defined as a term expressing the probability that during a year an individual will die as a consequence of a particular accident. Social risk is defined by the relationship between the number of deaths and the probability that this number will be exceeded [1].

The Chemical Process Quantitative Risk Analysis (CPQRA) Method is one of the most sophisticated methods for the numerical determination of individual and social risks. Application of the CPQRA method requires considerable specialized training of the working team and is immensely time-consuming.

The Method presented in this paper has been designed by the International Atomic Energy Agency in co-operation with UNEP, UNIDO, WHO. It is easy, not time-consuming, and recommended for the approximate estimation of risk. The method is based on the experience of many experts and the findings of analyses conducted on actual accidents. The method is applicable for the risk assessment of major accidents with consequences beyond the boundary of the facility in the following cases [2]:

a) Fixed facilities where dangerous substances are processed, stored and handled.
b) Transportation of dangerous materials by road, rail, pipeline and water (not presented due to length limitation of this article)

DESCRIPTION OF THE METHOD AND PROCEDURE

A. First, it is necessary to define the monitored area, collect a list of all dangerous facilities and substances in the region and information on them. It is also essential to gather information on all the transport routes and the modes of transport for the dangerous substances.

K. Mahutova et al. (eds.), Defense and the Environment: Effective Scientific Communication, 167–176.

B. Second, the impact assessment (typically mortality) of potential accidents is carried out for each activity under consideration. Factors considered during this process include the affected area; density of population; distance and location of population and potential mitigating factors.

C. Third, the probability of occurrence of an accident is assessed based on the frequency of an accident occurrence for each activity being assessed. The procedure of probability estimation is based on the so-called probability figures, obtained from the study of accidents of analogous nature. Correction factors are then considered based on the frequency of decanting or filling, impact of the installed security systems (flammable materials only), safety and organisational measures, likely direction of wind, etc.

D. Fourth, social risk is assessed for each activity. All dangerous activities in the area are shown in a matrix, which represents the link between probability and consequences.

E. Finally, risk priorities are determined. Acceptability criteria of both individual and social risks have to be determined before the task is solved. The criteria are derived from national or international standards. The procedure reveals all the activities in the given area that are unacceptable or for which the risk must be reduced [2].

IMPACT ASSESSMENT

The following equation is used for the impact assessment of a particular activity in case of a fixed source:

$$\mathbf{C}_{i,s} = \mathbf{A} \cdot \mathbf{d} \cdot \mathbf{f}_A \cdot \mathbf{f}_m \qquad (1)$$

where $\mathbf{C}_{i,s}$ represents impact of an activity **i** and a substance **s**, **A** means the affected area in [ha], **d** means the density of population in the inhabited area being affected, $\mathbf{f}_A$ is the correction factor for the distribution of people in the affected area and $\mathbf{f}_m$ correction factor for the reduction of impact.

Table 1. Examples of the Record of Characteristic Substances

Accident reference number	Type of substance	Examples of chemical substances
1-3	Flammable liquid tension of vapour < 0,03 MPa at temperature 20 °C flash point > 20 °C	allyl-alcohol, acetaldehyde, acetone, aniline, benzene, butanol, dioxane, diesel oil, ethanol, ethylene glycol, ethylacetate, heptane, hexane, methanol, nitrobenzene, oil, phenol, pyridine, styrene, toluene, xylene
4 - 6	Flammable liquid tension of vapour > 0,03 MPa at temperature 20 °C	diethylether, petrol, liquid natural gas, propanol
7 – 9	Flammable gas liquefied by pressure	1,3-butadiene, butane, ethane, propane, propylene
10, 11	Flammable gas liquefied by cooling [a)]	liquid natural gas, methane
12	Compressed flammable gas	ethylene, hydrogen, methane, natural gas
13	Flammable gas in bottles	acetylene, butane, hydrogen, propane-butane (LPG), propane
14, 15	Explosives	ammonium nitrate (fertiliser type I), ammunition, nitroglycerine, trinitrotoluene
16, 17	Low-toxic liquids	chloropicrin, ethylisocyanate, phosphoric chloride, trichlorsilane, vinylidene chloride
18 – 21	Medium-toxic liquids	acrolein, acrylonitrile, methyliodide, methyltrichlorsilane, nitric acid, oleum, propyleneoxide, stannic chloride
22, 25	High-toxic liquids	hydrogen cyanide, tetroxide nitrogen, sulphur trioxide, tetrabutylamine
26, 29	Very high-toxic liquids	methylisocyanate
30, 35	Low-toxic gases	oxiran, vinylchloride
31, 36, 40	Medium-toxic gases	ammonia, boron trifluoride, carbon monoxide, hydrogen fluoride, silane, sulphur dioxide
32, 37, 41, 42	High-toxic gases	chlorine, formaldehyde, hydrogen bromide, hydrogen chloride, methylchloride, dioxide nitrogen
33, 38	Very high-toxic gases	dicyan, dimethylether, phosgene,
34, 39	Extremely toxic gases	hydrogen arsenide, ozone

[a)] See also record of flammable gases liquefied by pressure (accident reference number 7-9)

Table 2. Record of Referential Accidents–Classification of Substances According to Impact

Accident Reference Number	Type of Chemical Substance	Description of Substance Characteristics		Activity
1	Flammable liquid	Tension of vapour < 0,3 bar at temperature 20 °C		Storage in reservoir
2				Pipe-line
3				the others
4		Tension of vapour > 0,3 bar at temperature 20 °C		Storage in reservoir
5				Pipe-line
6				The others
7	Flammable gas	Liquefied by pressure		Rail, road, above ground reservoir
8				Pipe-line
9				The others
10		Liquefied by cooling		Storage in reservoir
11				the others
12		Under pressure		Pipe-line
13				Storage of pressure bottles (25–100 kg)
14	Explosive	Freely sprinkled		
15		Packaged (e.g. ammunition)		
16	Toxic liquid	Low-toxic		Storage in reservoir
17				The others
18		Medium-toxic		Storage in reservoir
19				Road/rail
20				Water
21				The others
22		High-toxic		Storage in reservoir
23				Road/rail
24				Water
25				The others
26		Very high-toxic		Storage in reservoir
27				Road/rail
28				Water
29				The others
30	Toxic gas	**Liquefied by pressure**	Low-toxic	
31			Medium-toxic	
32			High-toxic	
33			Very high-toxic	
34			Extremely-toxic	
35		**Liquefied by cooling**	Low-toxic	
36			Medium-toxic	
37			High-toxic	
38			Very high-toxic	
39		**Liquefied by cooling**	Extremely-toxic	
40		**In pipe-line**	Medium-toxic	
41			High-toxic	
42		High-toxic gasses Under pressure > 25 bar		
43		Toxic fumes of burning		from pesticides
44				from fertilisers (nitrogenous)
45				from sulphuric acid
46				from plastics (with chlorine)

Table 3. Determination of Impact Categories

Accident Reference Number	Quantity [t]								
	0,2-1	1-5	5-10	10-50	50-200	200-1000	1000-5000	5000-10000	> 10^4
1	- [a]	-	-	-	-	AI	BI	BI	CI
2	-	-	-	-	-	-	-	-	-
3	-	-	-	AI	BI	CI	DII	X [b]	X
4	-	-	-	-	-	BI	CII	CII	DII
5	-	-	-	-	-	-	-	-	-
6	-	-	-	BII	CII	DII	EII	X	X
7	-	AI	BI	CI	DI	EI	X	X	X
8	-	-	-	-	-	-	-	-	-
9	-	BII	CIII	CIII	DIII	EIII	X	X	X
10	-	-	-	-	-	BI	CII	CII	DII
11	-	-	-	BII	CII	DII	EII	X	X
12	-	-	-	-	-	-	-	-	-
13	-	-	CIII	CII	CI	CI	X	X	X
14	AI	BI	BI	CI	CI	DI	X	X	X
15	BIII	BIII	CIII	CI	CI	DI	X	X	X
16	-	-	-	-	-	AII	AII	BII	CIII
17	-	-	-	AIII	AII	BII	CII	CII	CII
18	-	-	-	AIII	BIII	DIII	EIII	FIII	FIII
19	-	AII	CIII	DIII	X	X	X	X	X
20	-	BII	DIII	EIII	FIII	GIII	X	X	X
21	-	BII	CIII	DIII	EIII	FIII	FIII	X	X
22	-	-	AII	BIII	CIII	EIII	FIII	GIII	GIII
23	BII	CII	DIII	EIII	X	X	X	X	X
24	CII	DII	EIII	FIII	GIII	GIII	X	X	X
25	BII	CII	DIII	EIII	FIII	FIII	GIII	X	X
26	AII	BII	CIII	EIII	FIII	GIII	GIII	HIII	HIII
27	CII	DIII	EIII	FIII	X	X	X	X	X
28	DIII	EIII	FIII	GIII	HIII	HIII	X	X	X
29	CIII	DIII	EIII	FIII	GIII	HIII	HIII	X	X
30	AII	BII	BII	CIII	CII	DIII	DIII	DIII	EIII
31	BII	CII	CII	DIII	EIII	FIII	FIII	GIII	HIII
32	CII	DIII	EIII	EIII	FIII	GIII	GIII	X	X
33	DIII	EIII	FIII	GIII	GIII	HIII	X	X	X
34	EIII	FIII	GIII	HIII	HIII	X	X	X	X
35	-	-	-	AII	AII	BII	BII	CII	DIII
36	-	AII	BII	CII	DIII	DIII	EIII	FIII	GIII
37	BII	CII	DIII	EIII	EIII	FIII	FIII	GIII	HIII
38	DIII	EIII	FIII	FIII	GIII	GIII	X	X	X
39	EIII	FIII	GIII	HIII	HIII	X	X	X	X
40	-	-	-	-	-	-	-	-	-
41	-	-	-	-	-	-	-	-	-
42	-	-	-	-	-	-	-	-	-
43	-	-	-	BII	DIII	EIII	EIII	X	X
44	-	AII	AII	CIII	EIII	FIII	FIII	X	X
45	-	-	AII	BII	CIII	DIII	DIII	X	X
46	-	-	-	AII	CIII	DIII	DIII	X	X

a) This symbol represents almost negligible impact

b) This symbol means such a combination substance-quantity, that does not occur in practice

Risk assessment procedure is as follows [3]:

a) Risk source is chosen.

b) Dangerous substance [a] is characterised according to Table 1.

c) Accident reference number is determined according to Table 2 with the information on the character of dangerous substance.

d) Categories of impact (AI - HIII) [b] are determined in compliance with Table 3, which is based on the information both on the quantity of the substance being assessed and the accident reference number.

e) Assessment of both the range of impact and the area being affected by accident is carried out according to Table 4.

f) Distribution of population is assessed in the area of a circular shape the radius R of which represents the maximum range of impact. Population density in the most significant parts of affected area is assessed. If the population density data is not available, assessment from Table 5 is to be used.

g) Correction factor f_A is found in Table 6 based on the proportion of inhabited area in the affected region.

h) The value of reduction factor f_m is found in Table 7 based on information on the characteristics of dangerous substance. [c]

i) The impact of a particular accident on inhabitants is assessed with the help of equation (1).

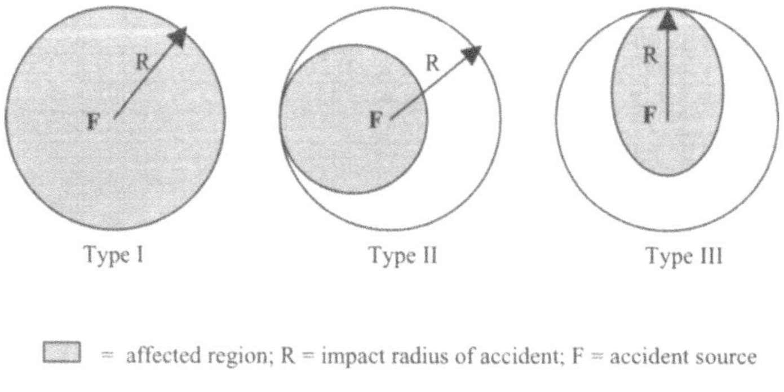

Figure 1. Shapes and Area of Affected Region Corresponding to the Impact Radius

---

a) If there is more than one substance used during the activity and each substance can cause an accident, then each case is considered separately. If a group of substances can impact together, one equivalent substance is considered (quantity sum). If a flammable substance is also toxic, then both cases have to be assessed. After several steps being taken it is clarified if the flammability is relevant in comparison to toxicity.

b) Distance of impact is increasing from A to H. Numbers I - III are related to the affected area and its shape and they are shown in Figure 1.

c) Possibility to warn inhabitants is respected.

Table 4. Maximum Range of Accident and the Area of Affected Region

Impact Categories	Range of Accident [m]	Area of Affected Region [ha]		
		Type I	Type II	Type III
A	0-25	0,2	0,1	0,02
B	25-50	0,8	0,4	0,1
C	50-100	3	1,5	0,3
D	100-200	12	6	1
E	200-500	80	40	8
F	500-1000	-	-	30
G	1000-3000	-	-	300
H	3000-10000	-	-	1000

Table 5. Density of Population

Area Characterisation	Density of Population in the Area [inhabitants. ha^{-1}]
Agricultural area	5
Individual houses	10
Village, quiet residential area	20
Residential area/built-up area	40
Busy residential area	80
City centre, shopping centre, block of flats	150

Table 6. Values of Correction Factor f_A

Type of Affected Area	Value of Factor	Proportion of Inhabited to the Affected Area [%]				
		100	50	20	10	5
I	f_A	1	0,5	0,2	0,1	0,05
II		1	1	0,4	0,2	0,1
III		1	1	1	1	1

Table 7. Values of Reduction Factor f_m

Dangerous Substances	Numbers of referential accident	Factor f_m
Flammables	1-12	1
Flammables	13	0,1
Explosives	14, 15	1
Toxic liquids	16-29, 43-46	0,05
Toxic gases	30-34, 40-42	0,1
Toxic gases	35-39	0,05

ACCIDENT PROBABILITY ASSESSMENT

Accident probability assessment is determined by the following equations:

$$N_{i,s} = -\log P_{i,s} \quad (2)$$

$$N_{i,s} = N_{i,s}^{*} + m_l + m_f + m_o + m_p \quad (3)$$

where $P_{i,s}$ represents frequency of accident occurrence per year for the given activity **i** and dangerous substance **s**, $N_{i,s}$ means relevant probability number for an activity **i** and dangerous substance **s**, $N_{i,s}^{*}$ corresponds to the average value of the probability number for an activity **i** and dangerous substance **s**, m_l is a correction factor according to the frequency of source decanting or filling, m_f is a correction factor based on the use of security systems for flammable substances, m_o represents correction considering character of organisational measurements and safety management, and m_p is a correction factor including the direction of wind and the proportion of inhabited area to affected area.

The procedure for the probability assessment of an accident occurrence is the following [3]:

a) Source of risk is chosen as in case of impact assessment.

b) Dangerous substance is characterised according to Table 1 [a)].

c) Accident reference number is determined according to Table 2 and information on the character of dangerous substance.

d) Average value of probability number $N_{i,s}^{*}$ is given to each dangerous substance identified for each activity using Table 8.

e) Value of correction factor m_l for the frequency of decanting or filling is determined with the help of Table 9.

f) Correction for flammable gases follows. The value of correction factor m_f is assessed from Table 10. It considers security systems being used and number of pressure bottles being stored.

g) Value of correction factor m_o is determined from Table 11. This parameter considers organisational and safety aspects, e.g. number of the operational hours of equipment, level of safety management, existence and quality of safety procedures, quality and experience of service and maintenance, elaborated emergency and evacuation plans, etc.

h) Correction factor m_p, which considers direction of wind and proportion of inhabited area to affected area, is assessed according to Table 12.

---

a) If the accident can be caused by more substances independent of each other, then each substance is considered separately. If the substances have the impact together they are considered as one equivalent substance.

i) Finally the probability $\mathbf{P}_{i,s}$ of the recognised accident is calculated according to equations (2) and (3).

Table 8. Probability Number Average Value $N_{i,s}^{*}$ for fixed facilities

Dangerous substance	Accident reference number	Activity	
		Storage	Process
Flammable liquid	1-3	8	7
Flammable liquid	4-6	7	6
Flammable gas	7	6	5
Flammable gas	9	7	6
Flammable gas	10, 11	6	-
Flammable gas	13	4	-
Explosives	14, 15	7	6
Toxic liquid	16-29	5	4
Toxic gas	30-34	6	5
Toxic gas	35-39	6	-
Toxic gas	42	5	4
Products of burning	43-46	3	-

Table 9. Correction of Probability Number for the Frequency of Decanting or Filling [a)]

Frequency of decanting or filling [per year]	Parameter m_l
1-10	0,5
10-50	0,0
50-200	- 1,0
200-500	-1,5
500-2000	-2,0

Table 10. Correction of Probability Number for Flammable Gases

Dangerous substance	Accident reference number	Safety measures	Number of pressure bottles in store	Parameter m_f
Flammable gas	7, 13	sprinkle system		0,5
Flammable gas	10	double casing		1,0
Flammable gas	13	fire wall		1,0
Flammable gas	13		5-50	1,0
			50-500	0,0
			> 500	- 1,0

a) Valid for all sources of risk except pipe-lines and pressure bottles.

Table 11. Correction of Probability Number for Safety Control [a)]

Operational Experience and Practice	Parameter m $_o$
Above-average	0,5
Average	0,0
Below-average	- 0,5
Bad	- 1,0
Non-existing	- 1,5

a) The following factors are included: safety management, number of years the system is in operation, maintenance, documentation and procedures, training, emergency planning, etc.

Table 12. Correction of Probability Number for the Direction of Wind in Relation to the Affected Region Inhabited Area

Category of affected area	Parameter	Affected region inhabited area [%]				
		100	50	20	10	5
I	m_p	0,0	0,0	0,0	0,0	0,0
II		0,0	0,5	0,5	0,5	0,5
III		0,0	0,5	0,5	1,0	1,5

CONCLUSION

Risk assessment is the fundamental prerequisite for successful risk management. An easy and not time-consuming method was presented. It is recommended for the approximate risk quantification for some military facilities. The method has to be used reasonably because it considers only some aspects. For a more accurate risk assessment especially with respect to the environmental impacts of major accidents a more accurate method should be used, e.g. Chemical Process Quantitative Risk Analysis Method (CPQRA) that represents the complex study of safety.

REFERENCES

[1] Bozek, F. Risk Management and Implementations for Sustainable Environmental Management. 45 s. In KRAMER, M. et al. *Grundlagen eines ganzheitlichen Umweltmanagements.* 1. Ed. Germany, Wiesbaden: Gabler Verlag. (In print.).

[2] International Atomic Energy Agency. *Manual for the Classification and Prioritisation of Risks Due to Major Accidents in Process and Related Industries.* Vienna: IAEA 1993. 125 p.

[3] Bozek, F. Novakova K. and Urban, R. *Risk Assessment of the Facilities Handling Dangerous Substances,* In Proceedings of the International Conference on Crises Management in the Specific Surroundings, Zilina: University of Zilina,, 2002, 22 - 23 May, Part II, p. 53 - 58.

Risk Assessment of Large Capacity Petrol Storehouse

František Božek, Aleš Komár and Josef Navrátil
Military University of the Ground Forces, Faculty of State Defence Economy and Logistic Vyskov, Czech Republic

INTRODUCTION

A large capacity military storehouse for petrol is a typical military facility handling dangerous substances (flammable compounds). In our case petrol is stored in six subsurface storage tanks of a cylindrical shape with the total capacity of 100 m^3 of each storage tank. The storage tanks are filled up to 95 % of their maximum volume. Decanting is carried out 160 times a year and each tank is re-filled four times a year on the average. The storehouse is located at the edge of army barracks and workshops are placed 30 m south of the storehouse. Twenty-one personnel per hectare work at workshops on the average.

RESULTS AND DISCUSSION

A method developed by the International Atomic Energy Agency [1] was used for the risk assessment of the above mentioned facility.

First, it is necessary to find out the total amount **m** of petrol in tons being stored [2]. The amount is calculated according to formula (1):

$$\mathbf{m} = \mathbf{n} \cdot \mathbf{K} \cdot \mathbf{V} \cdot \mathbf{d} \cdot \mathbf{10^{-5}} \qquad (1)$$

where **n** is number of storage tanks, **K** means maximum filling capacity of storage tank in percentages, **V** represents capacity of storage tank in [m^3] and **d** = 710 is the density of petroleum in [$kg.m^{-3}$].

After the introduction of particular figures into the equation (1) it is possible to obtain the quantity of storage petrol:

$$\mathbf{m} = 6 \cdot 95 \cdot 100 \cdot 710 \cdot 10^{-5} = 404{,}7 \text{ tons of petrol} \qquad (2)$$

K. Mahutova et al. (eds.), Defense and the Environment: Effective Scientific Communication, 177–180.

IMPACT ASSESSMENT

In the Table of record of characteristic substances it is easy to find that petrol belongs to flammable liquids [1] with tension of its vapour higher than 0,03 MPa at temperature 20 °C. With regard to the fact that petrol is stored in subsurface storage tanks it has referential accident No 4 (see the Table of record of referential accidents). BI category of impact corresponds to the above mentioned quantity [equation (2)] of petrol according to the table of impact categories and their determination. If there was an accident, its range of impact would be 25 - 50 m and the affected area would be approximately 0,8 ha according to the table describing maximum range of accident and the area of affected region.

With regard to the fact that the number of staff in the working area is 21 per hectare and the proportion of the working area is assessed to be 10 %, the impact assessment $\mathbf{C}_{i,s}$ is based on equation (3) as follows:

$$\mathbf{C}_{i,s} = \mathbf{A} \cdot \mathbf{d} \cdot \mathbf{f}_A \cdot \mathbf{f}_m = 0{,}8 \cdot 21 \cdot 0{,}1 \cdot 1 = 1{,}68 \qquad (3)$$

Values of both f_A and f_m factors were found out from the tables of correction factor f_A values and reduction factor f_m values.

ACCIDENT PROBABILITY ASSESSMENT

The value of probability number for this case can be obtained provided that the number of referential accident is four and at the same time adequate tables are respected. That means tables of probability number average value $N_{i,s}^*$ for fixed facilities, of probability number correction for frequency of decanting or filling, of probability number correction for safety control and finally of probability number correction for the direction of wind in relation to the affected region inhabited area. The calculation was provided according to equation (4). It is based on the frequency of refilling and decanting of petrol storage tanks, acceptance of safety measures, service staff experience and proportion of the affected region working area to a total area:

$$\mathbf{N}_{i,s} = \mathbf{N}_{i,s}^* + \mathbf{m}_l + \mathbf{m}_f + \mathbf{m}_o + \mathbf{m}_p = 7 + (-1{,}0) + 0{,}0 + 0{,}5 + 0{,}0 = 6{,}5 \quad (4)$$

With the use of relation (5) the assessment of accident frequency $\mathbf{P}_{i,s}$ is obtained for this facility:

$$\mathbf{N}_{i,s} = -\log \mathbf{P}_{i,s} = 6{,}5 \Longrightarrow \mathbf{P}_{i,s} = 3{,}1 \cdot 10^{-7} \qquad (5)$$

Attained values were used for the individual and social risk assessment studied military storehouse. The situation is shown in Figures 1 and 2.

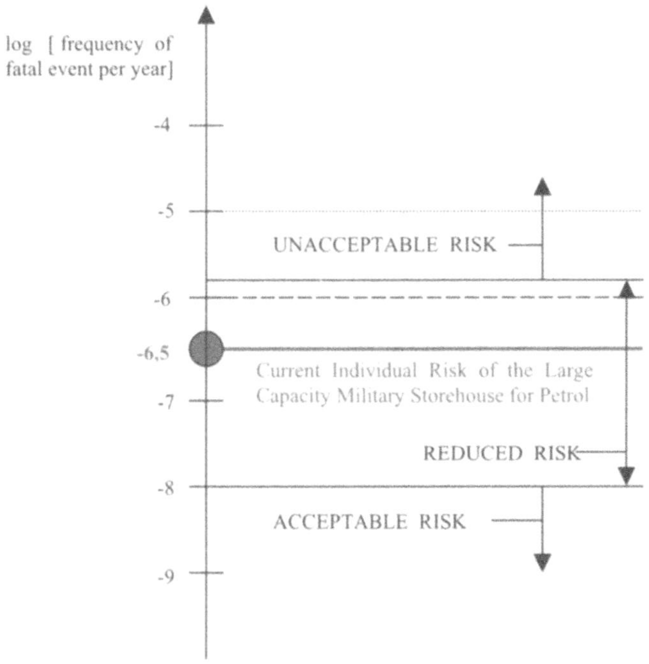

Figure 1 Individual Risk Acceptability Level for Studied Source – Storehouse for Petrol

Legend —— standards of the some EU countries; .... standards of the Czech Republic for existing facilities; – – – standards of the Czech Republic for the new facilities; ▬ current individual risk of the large capacity military storehouse for petrol

CONCLUSION

Risk assessment is the fundamental prerequisite for successful risk management. By the application of international standards on risk acceptability, it can be stated that social risk resulting from the studied source is acceptable. Individual risk significantly exceeds level of acceptance and it should be reduced, e.g., at the source or by successive transfer of workshops further from the source of risk. In our opinion it would be very purposeful to compare gained results with similar data which were estimated by other risk assessment procedures and verify the above described simple and not time consuming method for practical usage in the military sector.

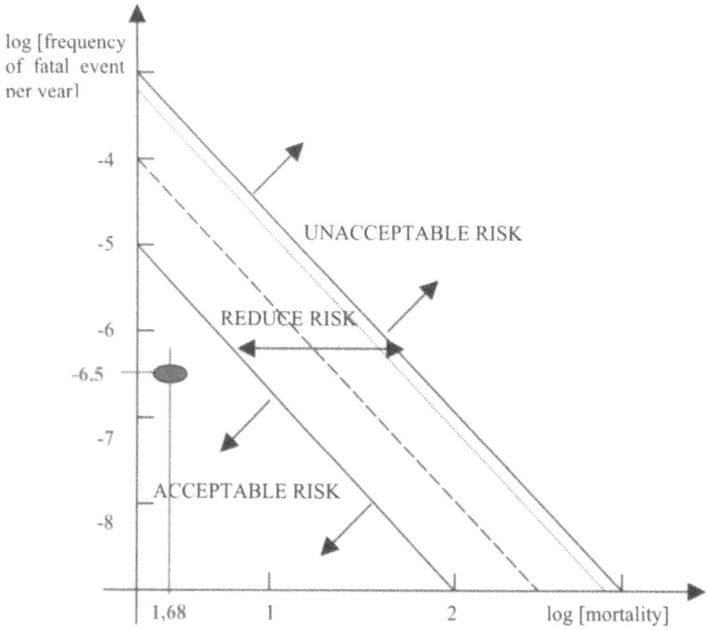

Figure 2 Social Risk Acceptability Level for Studied Source – Storehouse for Petrol

—— standards of the some EU countries; ········· standards of the Czech Republic for existing facilities; – – – standards of the Czech Republic for the new facilities; ⬬ - current social risk of the large capacity military storehouse for petrol

REFERENCES

[1] International Atomic Energy Agency. *Manual for the Classification and Prioritisation of Risks Due to Major Accidents in Process and Related Industries.* Vienna: IAEA 1993. 125 p.

[2] Bozek, F. Novakova K. and Urban, R. *Risk Assessment of the Facilities Handling Dangerous Substances,* In Proceedings of the International Conference on Crises Management in the Specific Surroundings, Zilina: University of Zilina,, 2002, 22 - 23 May, Part II, p. 53 - 58.

The Roving Windows Methodology for Risk Management and Remedy Optimization, An Approach for Addressing Chemical Contamination across Large Areas of Soil

Gaynor Dawson and Nathan Smith
Project Performance Corporation, West Richland, WA USA

INTRODUCTION

This paper describes the Roving Windows Methodology (RWM), a new software program developed as a customization to existing geographic information systems (GIS) architecture. The program is applied to large soil areas found to contain varying concentrations of hazardous chemicals where it serves both as a tool for risk assessment and as an aid to risk management. In the former mode, it is used to determine the risk to human health and ecological receptors posed by residual contaminant levels; while in the latter mode, it is used to identify the optimum volume of soil to be removed to achieve a target level of risk. In addition, RWM can be used to generate graphical presentations that help articulate the risk management options when clearly defined target risk criteria are not available.

BACKGROUND

Well-established guidelines exist for determining the risk posed by residual chemical concentrations in environmental media (EPA, 1999; EPA, 2001). Risk is assessed by calculating the dose a receptor will receive under a given scenario and then identifying the impact that would be expected from a dose of that magnitude. Dose predictions are made by taking the product of the mass or volume of contaminated media that is contacted (i.e., ingested, inhaled, or exposed to the skin) and the average concentration of contaminant in that media. Standard scenarios have been devised for use in selecting the former multiplier e.g., volume of water consumed per day, typical mass of soil ingested from contact on hands. Theoretically, the second factor, average concentration, should be readily determined through statistical analysis of site data. However, the latter requires designation of the boundaries within which the contact will occur. The boundaries are typically designated the exposure unit.

K. Mahutova et al. (eds.), Defense and the Environment: Effective Scientific Communication, 181–187.

In calculating human health risk, the exposure unit is defined by the nature of the activity during which the contact will occur. For instance, if contact is the result of a child eating soil while playing in the yard, the exposure unit is defined as a quarter- or half-acre back yard. If the receptor were a grounds worker, the exposure unit would be several acres depending on the size of operations and the nature of the worker's responsibilities.

In calculating ecological risk, the exposure unit can be equated to the home range of the receptor species that has been selected as the valued ecological resource. Smaller mammals such as mice may be associated with exposure units of one half acre. Larger receptors (e.g., robins, fox, deer) have larger home ranges and therefore, larger exposure units. As a consequence, for contaminants that do not bioaccumulate, risk is often driven by smaller receptors (i.e., the receptor with the smallest home range). For contaminants that do bioaccumulate, the risk is driven by effects to the top of the food chain, which are typically the larger species.

PROBLEM

At large sites with contaminated soil, there are virtually an infinite number of unique exposure units that could be delineated from within the overall site. Faced with this complexity and the need to make a compelling argument that all possible exposure units posing an unacceptable risk have been remedied, decision makers have often opted to remove or remediate every location from which a sample was observed with a concentration in excess of the risk-based criterion. This conservative approach results in remediation of a much larger volume of soil than is necessary and in so doing, raises the cost of remediation unnecessarily.

SOLUTION

A software program using the Roving Windows Methodology has been developed to allow for delineation of probable exposure units within large areas of contaminated soil. The program is designed as a module to be used in conjunction with a licensed version of ESRI's ArcView™ and Spatial Analyst Extension™.

The RWM application allows the user to quickly and easily develop assessments of various contamination cleanup scenarios and their effect on exposure levels associated with residual soil contamination. By examining the mean contaminant concentrations across an exposure unit, the tool provides an estimate for the level of exposure that is expected to occur at each point in the area of analysis under different sets of conditions. Subsequently, various cleanup options can be tested to determine their net effect on the probable exposure levels.

The application uses point contamination data to interpolate a continuous contamination surface across the study area. The Inverse Distance Weighted (IDW) method is the default algorithm used by the application to interpolate point data to a continuous grid. Spline or Kriging may also be used for interpolation if there are a sufficient number of data points or if there is sufficient information about the statistical relationship among points. The output contamination surface is a raster grid in which each grid square contains a unique contaminant concentration.

In cases where known site features limit or define the exposure to contaminated soil, RWM can modify the contamination grid to account for the effect of those features. For instance, if an asphalt cap is in place over a portion of the site such that it prevents human exposure to contaminants, RWM can remove that portion of the site from the analysis. Conversely, if features such as contaminated ditches account for all of the exposure risk for a site worker (i.e., the worker is responsible for keeping ditches clean) or ecological receptor (i.e., the ditch is the preferred habitat), RWM can remove all other areas of the site from the analysis and focus only on the ditches.

Using the contaminant concentration grid, the application provides an estimate of the level of exposure to contaminants at every grid cell in the study area. This estimate is calculated using a rectangular neighborhood window that is passed over each cell in the contamination grid. This neighborhood window combines grid cells in the contamination grid into a group to form the exposure unit (Figure 1).

The user can set the exposure unit to any size, the only limitation is the grid cell size (there should be at least 9 grid cells within an exposure unit). As the center of the exposure window passes over a grid cell, it calculates the average concentration of that cell and all surrounding cells that fall within the window. This average concentration is outputted to an individual cell in a new raster grid that corresponds to the center cell within the exposure window. This process is repeated as the center of the exposure widow passes over each cell in the contamination grid. The resulting output grid contains exposure values at each individual grid cell. After calculating the exposure levels across the study area, the RWM application highlights areas that exceed the target exposure level for the site.

The RWM application simulates soil removal actions with the goal of eliminating the contaminated grid cells that exceed the user-selected action level in areas where there is an unacceptable risk of exposure. Contaminated grid cells are removed (their contamination level is set either to zero or to the background level for the site) in areas where the exposure exceedence occurs.

The application allows the user to test various action levels to determine the optimal cleanup response in order to reduce exposure. By iteratively removing contamination grid cells at various action levels and re-running the exposure analysis to determine the impact of the removal action, the application determines the target cleanup level that eliminates unacceptable exposure risk.

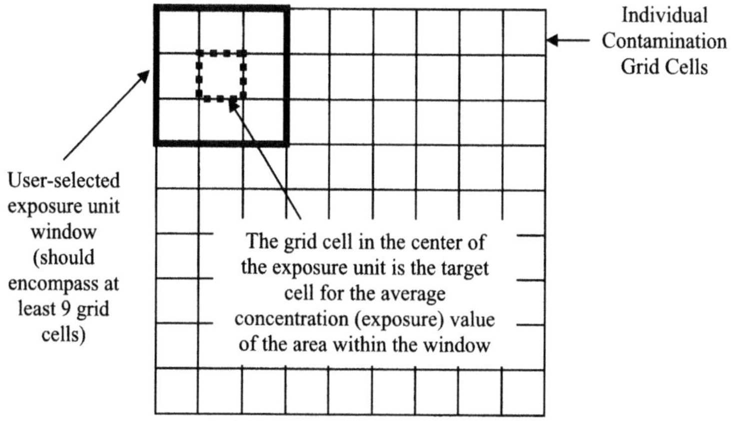

Input: A grid of contamination concentrations provides the input for the roving windows operation.
Process: The center of the exposure unit window is passed over each cell in the contamination grid. The average concentration of all grid cells that fall within the exposure window is calculated each time this occurs.
Output: Each time the center of the window passes over a cell, the average concentration within the exposure unit is outputted to a single cell in a new grid. The resulting grid contains exposure values in each cell across the study area.

Figure 1. Groupings of Grid Cells Make Up the Exposure Unit

Optimum target cleanup levels determined by this iterative process are generally higher than the target or threshold exposure level. This phenomenon occurs because areas of unacceptable exposure are often caused by hotspots of unusually high contaminant concentrations. In many cases, these hotspots are surrounded by soils with low levels of contamination, such that the overall average concentration within the exposure unit does not exceed the risk based exposure level. By differentiating those hotspots of contamination that result in unacceptable exposure risk from those that do not, RWM allows site managers to develop a focused and effective remedial response strategy.

Ecological risk assessment is complicated by the diversity of receptors that may be considered. An analysis of the literature provides for an array of information indicating the types of impacts that can result from contact with different concentrations of contaminants. Just as various species have evolved with greater and lesser tolerance for different conditions (e.g., fresh water versus salt water), so too different species are more or less resistant to different doses of chemicals. Some species even thrive on exposure to chemicals (consider the biota that degrade contaminants preferentially). As a consequence, there are no clear criteria for target cleanup levels with respect to ecological receptors. Rather, the risk manager is asked to determine what ecological resources are valued (e.g., species or biomes) and at what level they are to be protected (e.g., population or community level impacts as opposed to individuals within a species). Once those decisions have been made, a target criterion can be selected.

The RWM can be applied to assist the risk manager in selecting cleanup criteria. In this mode, the optimization program is run using a range of values for the target cleanup criterion. The output of these sequential runs is then portrayed in a graphical presentation that depicts the relationship between the cleanup criterion and the volume of soil that would have to be addressed (or the cost of remediation, since application

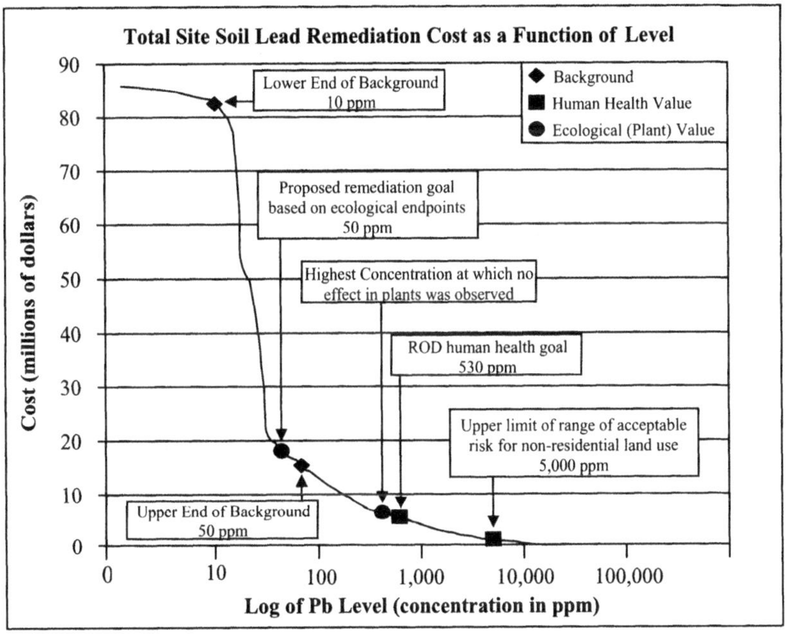

Figure 2. Example of Roving Windows Output Used to Assist in Ecological Risk Management

of a unit price for remediation quickly converts volume of soil to total cost). The relationship is then annotated with information taken from the literature indicating the nature of effects one would expect to see with exposure to a given concentration (Figure 2). By constructing such a graphical presentation, it is possible to put the cost-risk-benefit proposition in front of the decision maker in a very concise form.

EXAMPLE APPLICATIONS

The RWM has been successfully applied to two sites where soil contamination was thought to pose an unacceptable risk of exposure. RWM was used to evaluate exposure risk and to estimate the required response strategy for lead contaminated soil at a weapons development facility (WDF) and an ammunition production plant (APP). In both instances the RWM provided an estimated volume of soil to be removed that was smaller than the full extent of the area with unacceptable exposure risk.

Historic activities at the WDF resulted in areas of lead contaminated soil at levels that were thought to exceed risk-based exposure levels. The facility did not have a methodology in place for assessing the full extent of the exposure risk or an estimate of the amount of soil that would require removal/treatment in order to meet exposure levels. Existing soil sampling data provided the required input to the application, and were used to create a continuous contamination grid across the area of concern.

Based on this contamination surface, the total area of surface soil that exceeded the exposure limit within the area of concern was approximately 38,719 ft^2. This area of exposure, however, resulted from a localized point of contamination that was much higher than the surrounding contamination levels. The RWM application revealed that a 14,519 ft^2 area of soil with high levels of lead contamination was the cause of this unacceptable exposure risk.

Applied to the soil lead contamination issue at the WDF, the RWM application provided a clear analysis of the extent of areas that caused exposure to lead to an unacceptable level of risk and determined the scope of the required response action. In this scenario, RWM indicated that the desired risk level could be achieved while removing a 37.5% smaller volume of soil than would have been removed if all soil above the target average exposure concentration had been removed. Removing only those areas of soil that caused the unacceptable risk, rather than removing all soil with concentrations in excess of the target level saved a substantial amount of money while still achieving protectiveness.

Lead contamination in the soil at the APP was known to pose an unacceptable risk of exposure at five areas of concern. While the presence of unacceptable exposure at those areas was known, the full extent of the exposure risk under different sizes of exposure units, and the volume of contaminated soil that required removal were undetermined. The RWM application was used at this site to evaluate the extent of the exposure risk under various exposure unit sizes and to estimate the required remedial response to eliminate the risk.

In this analysis, two exposure unit sizes (5,000 ft^2 and 2 acres) were used to estimate the risk and the required response. The two exposure units represented residential use (5,000 ft^2) and industrial use (2 acres). The areas of concern that were examined in this analysis were bounded by known features (such as ditches and fences), so all soils that fell outside of those features were eliminated from the analysis. Because the resulting areas of exposure were all generally smaller than 1 acre, the exposure unit size was not expected to have a large impact on the extent of exposure risk or on the amount of soil required for removal. The results of the analysis show that under the 5,000 ft^2 exposure unit the required soil removal volume was 21,484 yd^2. Using the 2 acre exposure unit the estimated volume of soil to be removed was 21,059 yd^2. This 2% difference in removal volumes confirmed that exposure unit size was not significantly influential on exposure risk at this facility.

RWM was applied to the five areas of concern at the AAP using the 5,000 ft^2 exposure unit at depths of 0, 2, and 5 feet. The exposure risk and remedial response analysis was performed individually at each of these three depths because sampling data indicated that there were distinctly different levels of contamination at each depth (with the highest level of contamination at the surface and the lowest at the 5 foot depth). The results of the three analyses were combined to provide a realistic estimate of exposure and soil removal volume down to five feet. No unacceptable exposure was found at the five foot depth in several of the areas of concern. In those areas, soil removal was only simulated between 0 and 2 feet.

The results of the APP analysis revealed that the total volume of soil that would be removed if all soil containing lead at a concentration greater than the target action level were removed would be 30,278 yd^3. However, if only the areas with lead concentrations high enough to drive the average across an exposure unit (based on the 5,000 ft^2 exposure unit) are removed, the required soil removal volume would be approximately 24,223 yd^3. This 20% reduction in soil removal volume represented a significant optimization of the response strategy achieved using RWM to locate the areas of contamination that caused the unacceptable exposure risk.

While these two example applications involved lead contamination in soils, RWM can be used to estimate exposure and response levels associated with any type of soil contamination. Regardless of the contaminant, the only required input to the application is a GIS file that contains point sampling data.

REFERENCES

USEPA 1999. Issuance of Final Guidance: Ecological Risk Assessment and Risk Management Principles for Superfund Sites. OSWER Directive 9285.7-28P. U.S. Environmental Protection Agency. 2001. Risk Assessment Guidance for Superfund:. Volume 1 Human Health Evaluation Manual. Office of Emergency and Remedial Response. Washington, D.C. OSWER Directive No. 9285-47. December

Environmental Technologies Research and Deployments at Los Alamos National Laboratory

Dianne Williams Wilburn
Los Alamos National Laboratory, Los Alamos, NM, USA

HISTORY OF LOS ALAMOS NATIONAL LABORATORY

In the summer of 1942, the United States was fully engaged in the prospect of a long war. A group of theorists, under the direction of J. Robert Oppenheimer, met at Berkeley during that summer to develop preliminary plans for designing and building a nuclear weapon. In early October 1942, General. Leslie Groves, commander of the Manhattan District of the Corps of Engineers, learned from scientists in the Office of Scientific Research and Development (OSRD) that they needed a new laboratory to concentrate their scattered theoretical and experimental efforts to design a nuclear weapon. Groves had security concerns and believed a remote site was required, which had a climate permitting year-round construction, provided safety from enemy attack, met specific transportation requirements along with access to adequate power, water, and fuel. The site would also have to provide an adequate testing ground and, for reasons of safety and security, need to be in a sparsely populated area. Oppenheimer proposed Los Alamos, New Mexico, whose primary occupants were the students and staff of the Los Alamos Ranch School, a residential boys' school. The school had a number of buildings suitable for housing scientists, and beautiful views from the mesa top to the east and of the Sangre de Cristo Mountains. Access to the mesa could easily be controlled by shutting the main entry road, and the road could be widened to accommodate trucks and heavy machinery. Canyons surrounding the site could be used for explosives tests. [1,2]

After World War II, the Atomic Energy Commission, predecessor to today's United States Department of Energy (DOE), and the University of California operated the laboratory to fulfill the nation's nuclear weapons mission. This mission came at price to the environment. The vast network of weapons facilities owned by DOE, located on thousands of square miles of Federal reservations in 13 States, has produced widespread contamination of the environment with toxic chemicals and radionuclides. [1]

K. Mahutova et al. (eds.), Defense and the Environment: Effective Scientific Communication, 189–196.

THE UNITED STATES DEPARTMENT OF ENERGY'S OFFICE OF ENVIRONMENTAL MANAGEMENT (EM)

There is inherent difficulty of establishing environmental cleanup priorities in an agency as complex as DOE. DOE has a variety of missions: national-defense, energy-security, environmental-quality, and basic-research in support of its other missions.

DOE developed a tool for setting environmental restoration priorities called the Environmental Restoration Priority System (ERPS), which was used from 1988 to 1991. However, the use of ERPS was discontinued, mainly due to strong opposition by states and other stakeholders who felt that the system had been developed without their input.[1]

During the 1980's, however, state and federal environmental regulators gained partial jurisdiction over DOE sites. Previously, DOE had considered itself to be largely responsible for its own environmental performance. Next, DOE's Office of Environmental Management, EM, was established in a 1989 reorganization. It was recognized that if demands of federal and state regulators were to be met and if activities associated with waste management and environmental restoration were increasingly complex; a centralized planning process was needed to take into account different environmental needs across the entire DOE complex.

LEGACY WASTE AND ENVIRONMENTAL SCIENCE NEEDS AT LOS ALAMOS NATIONAL LABORATORY [1]

To satisfy DOE as the owner, University of California as the operator, environmental regulators and stakeholders, Los Alamos National Laboratory must decide what actions to take and how to spend its resources for environmental cleanup. In making its decisions and undertaking its actions, priorities must be established.

Many of the processes used to carry out the Laboratory's past and present missions involved the use of hazardous and radioactive materials. During World War II, and for a while thereafter, some of these materials were disposed of on the Laboratory site or were otherwise released into the environment. During this time, the disposal of hazardous chemical and radioactive wastes was not regulated and therefore some of these materials were disposed of improperly.

The mission of the Environmental Restoration Project at Los Alamos National Laboratory is to remedy environmental problems caused by 50 years of Laboratory operations by bringing together multi-disciplinary, world-class science, and engineering. The Project's goals are to: protect human health and the environment from exposure to hazardous, radioactive, and mixed wastes from past treatment, storage, and disposal practices and meet the environmental cleanup requirements of the Laboratory's permit to operate hazardous waste facilities.

The Environmental Restoration Project originally identified over 2,100 potential release sites at and around the Laboratory. Contamination originated from septic tanks and lines, chemical storage areas, wastewater outfalls, landfills, incinerators, firing ranges and their impact areas, surface spills, and electric transformers. Potential release sites are found on mesa tops, in canyons, and in the Los Alamos townsite. As a result of the investigations, remediations, and the 1999 and 2000 Annual Unit Audits, the Laboratory has reduced the number of potential release sites requiring further action by over 60%. A small percentage of sites, currently estimated at less than 10 percent, will need to go through the entire corrective action process, a task that is expected to take until 2013 to complete. [1]

ENVIRONMENTAL SCIENCE AND TECHNOLOGY NEEDS [2]

RRES-ES program at Los Alamos has the responsibility for developing new and innovative environmental technologies, while enhancing older approaches that address actual environmental problems. We strive to enhance, develop, and focus the science and technology base of the Laboratory toward innovative developments and comprehensive solutions to environmental problems at the Laboratory as well as regional and global environmental problems. As a result, RRES-ES works with most segments of the environmental technology development community, including: other DOE sites; Department of Defense (DoD); Environmental Protection Agency (EPA); international organizations/agencies; many segments of private industry; several academic institutions; and others. Building upon the Los Alamos National Laboratory's history of excellence in multi-disciplinary problem solving, RRES-ES strives to be recognized as the leader in environmental technology development and implementation through scientific and engineering innovation, progressive management, quality facilities, exceptional business practices, and effective demonstrations and deployments.

The objectives of the RRES-ES Program are:

- development of an effective priority setting system, such as technology roadmaps, to prioritize technology needs;
- investment in basis research,, development, demonstration, and deployment of science and technology to enable successful site restoration; and

- integration of science and technology projects with the cleanup and operational programs at Los Alamos National Laboratory and across the DOE complex.

To establish Los Alamos science and technology needs, in FY01 an effective priority-setting system was developed at Los Alamos. Based on interviews and online input by project managers, the Laboratory listed 77 environmental science and technology needs necessary for site restoration. However, at that time, the RRES-ES program was involved in reorganization. Activities need to be completed to 1) match new technologies/technical responses with needs and provide detailed technical information to the environmental problem holders; 2) provide detailed technical information of the proposed technology choices to stakeholders, regulators and other involved parties, as appropriate; and 3) coordinate with vendors and technology developers to conduct new technology demonstrations at sites, as necessary. Thirty-three of the needs listed in FY01 were listed as high priority by project managers and are referenced in Table 1.

Table 1 – Los Alamos National Laboratory High Priority Environmental Science and Technology Needs

Need Title	**Waste Stream**
Real Time Tritium Surface Contamination Monitor	Low-Level Waste
Tritium Removal From Glovebox Atmosphere	Low-Level Waste
Cripper/sealer for tritium contaminated piping	Low-Level Waste
Large diameter Calorimeter	Low-Level Waste
Tritium waste compactor	Low-Level Waste
Characterization of tritium content in waste by measurement of helium3/helium4	Low-Level Waste
Portable detritiation system	Low-Level Waste
Improved PPE for tritium workers	Low-Level Waste
Improved technologies for D&D of tritium facilities	Low-Level Waste
Quality Assurance for TRU waste and associated costs	Transuranic Waste
Volume Reduction and Removal of Organic Constituents for Pu238 Cellulose Process Residue	Nuclear Materials
Volume Reduction and Removal of Organic Constituents for Pu^{238} Plastic/Rubber Process Residue	Nuclear Materials
Volume Reduction and Removal of Organic Constituents for Pu^{239} Cellulose Process Residue	Nuclear Materials
Volume Reduction and Removal of Organics Constituents for Pu^{239} Plastic/Rubber Process Residue	Nuclear Materials
Pu^{239} Metals and Compounds Flowsheet Development	Nuclear Materials
Pu^{239} Combustibles Flowsheet Diagram	Transuranic Waste
Pu^{239} Residues (Inorganic) Flowsheet Development	Nuclear Materials
Hydrogen Sensor System for Transportation and Storage Containers	Nuclear Materials
Hydrogen Getter for Transportation Containers	Nuclear Materials
Transport Characterization for Impure Plutonium Oxides and Residues	Nuclear Materials
Transportation Characterization for Impure Pu Metals and Alloys	Nuclear Materials

Need Title	Waste Stream
New Radiation Shielding Gloves	Nuclear Materials
Long-Term Gas Generation Surveillance	Nuclear Materials
Packaging for Transportation of Legacy Materials	Nuclear Materials
Integration of Smart Technologies into Nuclear Materials and Waste Storage	Nuclear Materials
Moisture Analysis Methods for Small Samples	Nuclear Materials
Moisture Analysis for Bulk Materials	Nuclear Materials
Efficient Separation of Thorium from Uranium	Nuclear Materials
Understanding Radiolytic Gas Generation in Pure Plutonium Oxides	Nuclear Materials
Understanding Radiolytic Gas Generation in Impure Plutonium Oxides	Nuclear Materials
Understanding Radiolytic Gas Generation in Plutonium Containing Residues	Nuclear Materials
Need for Sampling Arm and Ability to Obtain Representative Samples from the General's Tanks at TA-21 at Los Alamos National Laboratory	Mixed Low-Level Waste
Technology Need to Remove Sludge and Decontaminate Tank Interior at Los Alamos Radioactive Liquid Waste Treatment Facility at TA50: Tank TA 50-WM2-T5	Mixed Low-Level Waste

LOS ALAMOS NATIONAL LABORATORY RESULTS [1]

The RRES-SA program has completed 231 environmental projects since the inception of EM in 1989. Most projects have had an estimated life-cycle of 3 years. Projects are defined as research, development, demonstration and deployment projects. See Figure 1. Forty-six projects involved technology demonstrations, which provided further information for future deployments. Eighteen of these projects involved deployments of technologies for the first time for DOE environmental problems. Twenty-eight technologies have been transferred to the commercial sector following deployment and Laboratory staff have been awarded 4 R&D 100 awards. Five projects are discussed in this paper. Information concerning all projects is available: **http://www-emtd.lanl.gov**

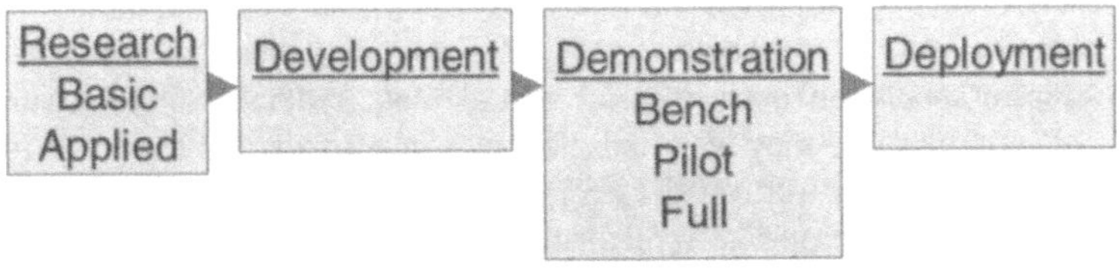

Figure 1 – Stages of Development from Research to Deployment.

Technologies have included Los Alamos expertise in the following areas:

- Actinide Stabilization, Destruction, and Cleanup
- Characterization, Monitoring, and Sensors
- Chemical and Physical Separation
- Cost Analysis
- High Level Waste Tank Characterization, Cleanup, and Safety
- Surface/Subsurface Contamination
- Characterization, Treatment and Disposal of Waste
- Robotics and Automation

RESEARCH AND DEVELOPMENT – ATMOSPHERIC PRESSURE PLASMA JET

The Atmospheric-Pressure Plasma Jet (APPJ), a non-thermal, atmospheric-pressure, glow-discharge plasma, can clean, decontaminate, etch, or coat surfaces at atmospheric pressure and low temperature. The APPJ technology enables use of plasma processing technology, normally a vacuum based method, at ordinary pressure, which greatly expands the applications and capability of this technology. By spraying surfaces in the open air, the plasma jet can convert a vast range of organic residues or toxins into water vapor, carbon dioxide, and other nontoxic gases or volatilize actinide materials such as plutonium and uranium for collection by filtration. Primary activities at Los Alamos National Laboratory include cleaning and detoxification of surface films on weapons components, decontamination of surrogate chemical and biological warfare agents, and etching and decontamination of actinide materials. The facility is also exploring applications for other materials ranging from microelectronics to building materials to health applications and treatment of textiles.

DEMONSTRATION - INTEGRATED SPECTROSCOPIC SYSTEM FOR ENVIRONMENTAL CONTAMINATION

Soil, sludge, and groundwater remediation methods are costly and inefficient. Although current analytical approaches can examine various chemical forms or species in the environment, these methods cannot address speciation over a broad spatial range. Metals are of particular concern because they can exist in multiple oxidation states. Without identifying each form of a contaminant before remediation begins, selecting the appropriate remediation strategy is done by trial and error. The Integrated Spectroscopic System for Environmental Contaminant Speciation (ISSECS) can improve the economics of remediation by providing accurate speciation data early in the process. ISSECS is a laboratory-based solution that uses a unique combination of four analytical techniques. Developed to address tough environmental analyses, such as soils contaminated with uranium or plutonium, ISSECS also can improve speciation of common industrial pollutants, including selenium, mercury, lead, and cyanide.

DEMONSTRATION - NON-TRADITIONAL IN SITU VITRIFICATION

Many Department of Energy (DOE) sites are contaminated due to various legacy operations. Containing radionuclides, heavy metals, organic materials, and buried wastes, these contaminants pose both human health and environmental risks if they are removed or if they are left in place. Traditionally, ex situ stabilization and disposal has been a primary method of remediation and restoration. Partially successful demonstrations utilizing top down in situ vitrification have offered a treatment option without physical removal. Los Alamos National Laboratory, along with MSE Technology Applications, Inc., and Geosafe Corporation (now AMEC), conducted an NTISV demonstration using planar geomelting, in which heat is applied to the subsurface. NTISV can reduce safety issues for waste that generates high gas volumes by heating a contaminated area from the subsurface, which allows gases to move to the surface where they are removed through an off-gas treatment system. This method creates tall and wide planar melts in the subsurface which are ideal for specific treatment zones and buried waste. Planar melts not only move up and down, but most importantly, sideways to connect areas of melts. The resulting product is an impermeable residual similar to obsidian, which eliminates the risk of contaminant migration.

DEPLOYMENT- MULTIPLE PERMEABLE REACTIVE BARRIER

Due to past legacy activities, Mortandad Canyon has the following contaminants of concern: Strontium, Plutonium, Americium, Nitrates and Perchlorates. In the fall of 2002, Los Alamos installed a permeable, reactive multi-barrier in the shallow alluvial groundwater system in the canyon. Barrier materials were selected based on development work at Los Alamos in a treatability study. By using a multi-layered barrier, a variety of contaminants can be removed, thereby preventing off-site migration.

DEPLOYMENT - MERCURY REMOVAL AND STABILIZATION BY POLYMER FILTRATION

Removal of mercury from mixed waste debris to the new universal treatment standard of 25 ppb is a formidable problem. A need exists for a decontamination and stabilization technology applicable to mercury contaminated debris, equipment, soils, and other materials at all DOE facilities. Los Alamos National Laboratory has developed a water-soluble polymeric decontamination system for removing both ionic and elemental mercury. This technology, called Polymer Filtration, applies soluble Hg-binding polymers to contaminated surfaces along with other dissolution-enhancing chemicals. The resulting metal-laden polymer leach solution is passed into an ultra-filtration chamber where polymer metal complex is concentrated, as the complex is too large to pass through the membrane. Mercury is stripped from the polymer and the polymer and leach solution are recycled for further debris leaching.

REFERENCE

1. "Why was Los Alamos Needed?" Fiftieth Anniversary Articles on the Web. http://www.lanl.gov/worldview/welcome/history/03_why-needed.html
2. Oppenheimer's Better Idea: Ranch School Becomes Arsenal of Democracy," Fiftieth Anniversary Articles on the Web. http://www.lanl.gov/worldview/welcome/history/07_school-arsenal.html
3. "Improving the Environment: An Evaluation of DOE's Environmental Management Program," Committee to Evaluate the Science, Engineering, and Health Basis of the Department of Energy's Environmental Management Program, 1995, National Academy of Sciences publication. http://bob.nap.edu/html/doeemp/
4. "Los Alamos National Laboratory Environmental Restoration Project" Los Alamos National Laboratory website. http://erproject.lanl.gov/about_us/who.html
5. "Environmental Management for Science and Technology, Strategic Plan," DOE publication, November 1998.
6. Los Alamos National Laboratory, Environmental Solutions Program Offices website: http://www-emtd.lanl.gov
7. Los Alamos National Laboratory, Environmental Solutions Program Offices website: http://www-emtd.lanl.gov

The Superfund Basic Research Program: A Model for Interdisciplinary Research to Reduce Risk and Prevent Exposure

K.C. Donnelly[1], W. Suk[2] and S.H. Safe[1].
[1]*Texas A&M University, College Station, TX, USA*
[2]*National Institute of Environmental Health Sciences, Research Triangle Park, NC, USA*

INTRODUCTION

Science and technology have brought many advances to our society. Physicians routinely remove and replace parts of human hearts; chemicals have been developed that can aggressively attack and destroy a wide variety of cancer cells; and, a number of diseases that were once thought incurable can now be treated. This rapid advance to our medical technology has been accompanied by a similar growth of industrial activities. In the United States, the early growth of industries also resulted in the uncontrolled release of a broad range of organic and inorganic chemicals into the environment. Many of the sites where these chemicals were released to soil, air or water in an uncontrolled fashion were designated as Superfund sites. Significant progress has been made in recent years to assess and remediate many of these sites in the United States. However, in certain parts of Eastern Europe, the identification and remediation of uncontrolled hazardous waste sites has not progressed as quickly.

The National Institute of Environmental Health Sciences established the Superfund Basic Research Program (SBRP) during the latter half of the 1980's to support basic research to improve methods for the assessment and remediation of contaminated sites. The assessment and remediation of an uncontrolled hazardous waste site represents a challenge that requires close collaborations between toxicologists, engineers and scientists from many disciplines. The complexities of these sites include the characteristics of the contaminant(s), both in terms of compound composition and concentration. In addition, contaminants may have been released into a variety of different media. Affected media at Superfund sites may include soils, air, groundwater, surface water and sediment. Contamination in groundwater may exist in both the dissolved phase, and a separate non-aqueous phase. The contamination of these media may also result in uptake of the contaminant by plants, or transfer

K. Mahutova et al. (eds.), Defense and the Environment: Effective Scientific Communication, 197–203.

of the contaminant into the food chain through livestock. Humans may become exposed to chemicals released from these sites through ingestion of contaminated drinking water, inhalation, or dermal contact. Although the release of a hazardous chemical into the environment may represent a concern, it does not represent a threat to human or ecological health unless it is determined that an exposure pathway has been completed. Health effects associated with chemical releases range from cancer to birth defects, reduced IQ and decreased fertility.

Ultimately, the concern at any of these sites is the effect(s) of exposure to these contaminants on human or ecological receptors. Ecological receptors may include flora and fauna, and may include local as well as migratory populations. Human exposure to chemicals released from hazardous waste sites may occur in an occupational setting, or may result from exposure to residential or recreational populations. For each exposure pathway, there can be dramatic differences in the dose and duration of exposures. In developing a methodology for assessing risk at these sites, regulators must consider exposures that can occur *in utero* and throughout the life of the receptor. The heterogeneity of the human population, in terms of lifestyle habits, genes, and nutrition, further complicates the risk assessment process.

The current SBRP includes 19 university-based programs. The list of programs includes:

- University of Washington
- Oregon Health Sciences Center
- University of California-Davis
- Univ. California-Berkeley
- Univ. of California-San Diego
- Univ. of Arizona
- Texas A&M University
- Univ. of Florida
- Univ. of Kentucky
- Univ. of Cincinnati
- Michigan State University
- Dartmouth College
- Boston University
- New York University
- Columbia University
- Mt. Sinai
- Harvard University
- Duke University
- Univ. of North Carolina-Chapel Hill

Each of these 19 programs includes several biomedical science research projects, and at least one non-biomedical science project. The major research categories in the current programs include mechanistic toxicology studies, remediation projects and biomarker research. A smaller number of projects study epidemiology, bioavailability and modeling. A broad range of chemical contaminants have been detected at Superfund sites in the United States. Most sites include complex mixtures of a variety of organic and inorganic chemicals. The most common chemical classes studied in the current SBRPs include metals, chlorinated organics, PCBs, polycyclic aromatic hydrocarbons, and complex mixtures. Research is also being conducted to investigate pesticides, volatile organic compounds, and halogenated aliphatic hydrocarbons. These chemicals may cause a range of diseases. The most common disease type investigated by current SBRPs is cancer. Research is also being conducted to investigate mechanisms of neurological disease, reproductive dysfunction, pulmonary/cardiovascular disease and others. The product of these programs is framework of basic and applied research that leads to information that can be used to improve health outcomes.

SCIENTIFIC FRAMEWORK

An important objective of the SBRP is the translation of research results into information that can be used to improve public health. A schematic diagram outlining this process is provided in Figure 1 (adapted from Olden, 2000). Molecular toxicology information can be used to determine how a chemical produces a toxic effect and to develop a dose-response relationship. These studies may lead to epidemiologic research that measures the frequency of an adverse effect in an exposed population. Studies may also be conducted to investigate biomarkers of exposure or to identify biomarkers that may preclude observation of an adverse effect. Another component of this model is research to identify sensitive genotypes. This research may serve to determine how over or under expression of certain genes may effect sensitivity to disease. For example, the over expression of certain cytochrome P450 enzymes may increase susceptibility to lung cancer due to increased activation of airborne carcinogens. Information from the basic research studies improves knowledge and understanding of the many factors that influence exposure and the risk of an adverse health effect. This information in turn can be used to improve the process of risk assessment; and, to develop better methods to manage or control risk. Finally, the outcome of the risk assessment and management process is to prevent disease or identify methods for promoting health in a population.

To promote public health, SBRPs also include a Community Education or Outreach Program. The translation of the results of basic and applied research into information that can be used to promote public health is an important component of the SBRP. These programs may deal with a specific community that has been affected by a Superfund site. Other Outreach programs deal with specific populations, such as delivery of health promotion information to students or health care professionals.

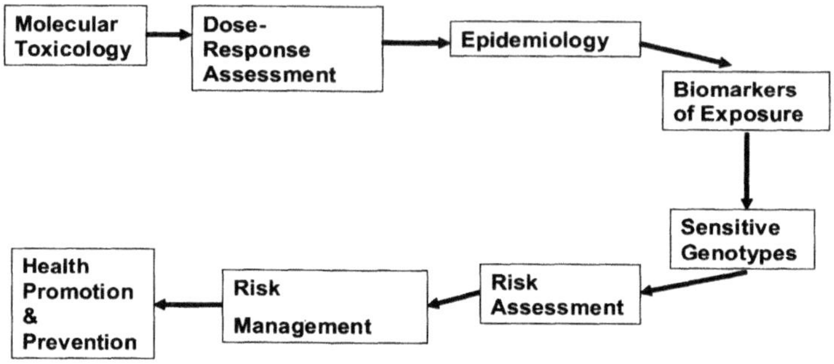

Figure 1. Schematic of process for translation of basic research results into health promotion and disease prevention programs.

Such educational programs provide a method for rapid delivery of current research results to potentially affected populations. These Outreach programs often provide valuable feedback to investigators as to the relevance of their research. In addition, they provide a resource for stakeholders that are concerned about the various factors associated with environmental health.

THE TEXAS A&M SUPERFUND BASIC RESEARCH PROGRAM

The Texas A&M SBRP consists of six research projects. The first project studies mechanisms of endocrine disruption. Many environmental chemicals have been shown to be capable of mimicking estrogen and other hormones in an exposed population. Project 1 studies both the mechanism and potency of these chemicals. The second project studies the genotoxic properties of complex mixtures. Current risk assessment procedures for complex mixtures often assume that the components of the mixture induce additive effects. By investigating the ability of model chemicals and isolated fractions from complex mixtures to damage DNA, a better understanding of the potential interactions of mixtures will be obtained. Project 3 is investigating the regulation of a gene in many different types of cells that appears to respond to stress. The fourth project is conducting studies with animals to evaluate genotypes that may have increased sensitivity to environmental chemicals. The fifth and sixth projects are non-biomedical science projects studying remedial techniques and exposure assessment. The fifth project is developing improved techniques for chemical treatment of contaminated media. Clay-based materials that have a high affinity for certain types of toxic chemicals are treated to increase their permeability. Field studies conducted as a part of Project 5 have established that the clays can be used to remove hydrocarbons and chlorinated phenols from a contaminated groundwater. The sixth project is conducting studies to measure both the efficacy of bioremediation, as well as an exposure assessment study. The Outreach Program at Texas A&M focuses on delivery of health promotion materials to high school students. In addition, courses have been offered in collaboration with the Texas Department of Health and other state agencies to health care professionals including nurses and medical students.

STRATIFIED EXPOSURE ASSESSMENT

A critical component of any study to predict adverse health effects from environmental chemicals is the exposure model. With the recent support of a US Environmental Protection Agency Science to Achieve Results (STAR) award, Texas A&M initiated a project to improve current models for exposure assessment. The goal of this project is to develop a stratified model for exposure assessment (Figure 2). Project 6 from the NIEHS SBRP is

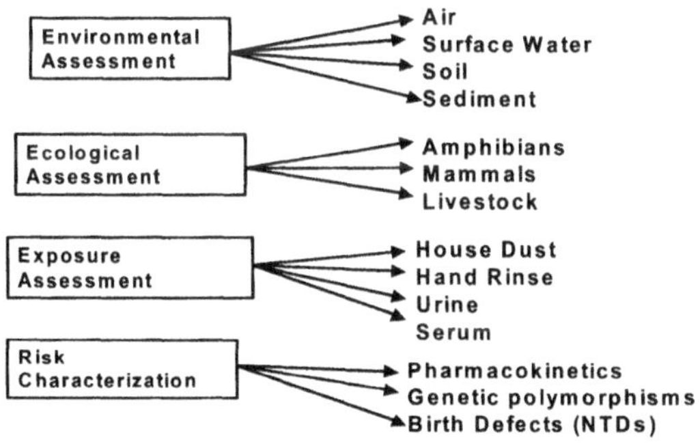

Figure 2. Methods used for improving existing models for exposure assessment.

conducting an environmental and ecological assessment. This portion of the study has collected environmental samples (including soil, surface water and sediment) from over 110 sampling stations in an industrialized area of Sumgayit, Azerbaijan. In addition, wildlife samples have been collected from many of these same areas for tissue analysis. These data will be used to determine if current exposure models are capable of predicting tissue levels of environmental contaminants. Wildlife are useful sentinels for exposure assessment because they are in contact with most environmental media, and because they do not have the confounding influence of lifestyle or occupational exposures.

The initial results from environmental monitoring have detected elevated concentrations of mercury, polycyclic aromatic hydrocarbons (PAHs), and chlorinated pesticides in soil samples. Analysis of tissue samples from frogs also detected elevated concentrations of chlorinated pesticides. Eight locations, three residential and five industrial were also used for collection of frog tissues. A minimum of nine frogs were collected from each location. Tissue levels of organochlorine pesticides were elevated at several locations near older industrial facilities, while concentrations in tissue samples from residential areas were not as high. The frequency of micronuclei damage was also measured in frog blood. Elevated frequencies of micronuclei in blood are an indication of gross damage to chromosomes. Although micronuclei were found to be significantly higher in blood from frogs collected near the wastewater treatment plant, there was not a direct correlation with contaminant concentrations in the environment.

The research to be conducted as a result of the STAR award will further investigate exposures in a refugee population living in areas adjacent to abandoned petrochemical facilities. A census is presently underway to provide information regarding the number and composition of families living near industrial facilities in Sumgayit. It is anticipated that 25 families, each including 2 children between the age of 2 and 10 years, 2 adults between the ages of 21 and 41, and 2 older adults over 51 years of age will be recruited. Control families will be recruited from the Caucas mountains in the northwestern region of the country. For the older children (over age 6) and adults, serum samples will be collected to measure concentrations of organic chemicals and, in isolated lymphocytes, DNA adducts. In addition, 1-hydroxypyrene levels will be measured in urine and hair samples collected to estimate exposure to mercury. Genetic polymorphisms will also be measured in blood collected from the refugees. When combined, this information will be used to evaluate exposure models.

Community meetings will be held at least once each year with these families. The purpose of these meetings will be to review the data from the previous year. In addition, families will be counseled as to possible methods that may be used to reduce environmental exposures. This may involve modifications to lifestyle exposures, such as smoking; or, in some cases, families may be encouraged to revise agronomic practices. Current information suggests that in several cases family gardens have been placed in areas where soil contamination may affect the quality of produce. A major objective of the SBRP is to prevent disease by reducing exposure. Efforts will be made in the current study to use the results from environmental, ecological and biomarker sampling to identify major sources of exposure and determine effective means of reducing these exposures.

CONCLUSIONS

The SBRP at Texas A&M combines the efforts of toxicologists, engineers, analytical chemists and wildlife ecologists. This interdisciplinary program is developing a broad approach to understanding the mechanisms of chemical toxicity, as well as the impact of environmental contamination on human and ecological health. Studies conducted at a remote location are very difficult. However, the ecological and human health studies conducted in this research offer a unique opportunity for such a comprehensive investigation. The initial studies have detected widespread and highly variable contamination. Ecological data indicate that amphibians may be exhibiting genotoxic effects as a result of their exposures. Studies have been initiated to investigate these exposures in a human population with a goal of identifying a means of reducing these exposures.

Primary prevention is a goal of public health. Primary prevention, simply stated, means that a disease has been prevented before it occurs. Children in the United States are routinely vaccinated for measles, mumps and rubella. This vaccination program has significantly reduced the incidence of these severe diseases in children. In a similar vein, interdisciplinary research programs have a goal of translating research results into information that can prevent disease or exposures. The National Institute of Environmental Sciences Superfund Basic Research Program and US Environmental Protection Agency Science to Achieve Results Program are designed to produce quality scientific knowledge that can be used to accomplish this goal. The complex interactions of chemical composition, multiple pathways of exposure, and genetic sensitivities make it difficult to accurately assess and manage risk. Investigations into these areas are more effective if scientists from multiple disciplines are involved. The outcome of these investigations may be information that can be used to improve the process of risk assessment through more accurate exposure information or data that defines sensitive populations. Studies have also been conducted to develop improved methods for detoxification of contaminated media. However, the ultimate goal of the SBRP is to translate research results into improved health outcomes. It has been suggested that when lifestyle and genetic factors are included, environmental factors contribute to as much as 90% of all disease. Interdisciplinary research programs such as the NIEHS SBRP or USEPA STAR foster results from basic research that can be translated into improved environmental health. Through the translation of research results into programs that can identify factors causing disease or improve methods for avoiding exposures, health education and promotion is an important metric of the success of these programs.

REFERENCES

1. Olden, K. 2000. Presentation at NIEHS Annual Superfund Basic Research Program meeting, Chapel Hill, North Carolina.

PART V:

PROTECTION OF NATURE AND NATURAL SYSTEMS

PROTECTION OF RARE SPECIES AND HABITATS IN THE ARMED FORCES OF LITHUANIA

Irma Jakimavičiūtė
Ministry of National Defense of Lithuania, Vilnius, Lithuania

INTRODUCTION

Lithuania has ratified the Convention on Biological Diversity and undertook obligations to control use and protection of biological resources, identify the most important components of biodiversity, carry out their monitoring and regulate the activity likely to have a negative impact. Currently 5 laws regulate protection of natural resources in Lithuania, a lot of secondary legislation (standards, lists, rules, recommendations etc.) assists in turning law requirements into action. A rather efficient system of protected territories (11.9 % of the Lithuanian territory) plays an important role in preserving biological diversity. The governing document for the biodiversity protection related work is the Strategy for the Conservation of Biological Diversity of the Republic of Lithuania (1998).

In order to meet national goals regarding the protection of biological diversity and those set by international agreements and EU regulations the Ministry of National Defence included an objective "to plan military activities such that adverse impact on biodiversity and forests is mitigated" into the Defence Environmental Strategy (2001) and Program (2002). The objective committed the Ministry of National Defence to develop and maintain a GIS Biodiversity Database as well as military training areas and to develop a system for protection, maintenance and observation of biodiversity, especially for species at risk in the training areas.

NATURE MANAGEMENT IN MILITARY TRAINING AREAS

The main challenge that must be considered and solved is an efficient combination of nature conservation and military training. The best solution is a simple and workable nature management plan with fully coordinated military and ecological requirements. Both environmental

K. Mahutova et al. (eds.), Defense and the Environment: Effective Scientific Communication, 207–216.

organizations, mostly interested in proper conservation measures, and military organizations, the main executers of these measures, must be aware of all details of the plan and constantly coordinate the plan execution.

Nature management planning is a four step process. The first step is the inventory of rare species, communities and habitats. The inventory is carried out by scientific experts and the quality of data must be comparable to and match the criteria of investigations in the adjoining civil areas. The second step is the development of a nature management plan including a) identification of sites of great natural importance according to the criteria, which agree with ones used in civil territories, b) planning of conservation measures in line with assessment of sites of great natural importance and in regard to the influence of military land utilization, and c) development of a timely action plan. The third step is the implementation of the plan. And the fourth step is monitoring and revision of the plan. Close cooperation between military and scientific experts is essential during all steps of nature management planning.

The mentioned nature management planning approach was tested as a pilot project while developing the nature management plan for the Central Training Area of the Lithuanian Armed Forces.

NATURE MANAGEMENT PLAN OF THE CENTRAL TRAINING AREA

The Central Training Area was established in the beginning of the 20th century. The Training Area covers 2,220.1 ha and is surrounded by, so called, military training territory (buffer zone), which size is 6,265.89 ha and is owned and governed by the Forestry Service. The Training Area is located in Pabradė forest, on the right bank of the Žeimena River on the South-eastern sandy flat plain dominated by infertile sandy soils and an undulating ravine landscape. There are 6 lakes (total area 37.2 ha), many boggy areas and 2 small streams in the Training Area territory. The following biotopes can be distinguished in the territory: open sandy areas (inland dunes, heath lands), forests (dominated by pine trees), bogs (raised bogs around forest lakes and transitional bogs around lakes and fens) and meadows (mostly natural).

Inventory. The inventory of rare species, communities and habitats was carried out by experts of plants, communities, molluscs, insects, amphibians-reptiles, birds and mammals from the Institutes of Botany and Ecology, as well as from a couple of non-governmental organisations dealing with biodiversity conservation in the Central Training Area in May-October 2001. Data collected during former investigations (in 1993 and 2000) were also employed. During investigations rare plant and animal species, pant communities and habitats were examined in terms of diversity and state of their populations. Localities of rare species, localisation of protected plant communities and valuable habitats were identified using GPS (Geographic Positioning System) device. Boundaries of identified sites of great natural importance (with the highest concentration of rare species, communities and valuable habitats) were marked on the map.

Subsequent to the inventory results for the Central Training Area, is considered as a highly valuable territory in terms of nature, where due to specific training area activities, many rare and vanishing species have found refuge and a number of rare and interesting habitats formed over the years. 74 species (27 plants, 10 insects, 3 amphibians, 28 birds and 6 mammals), 1 plant community and 10 habitats included in the Lithuanian Red Data Book, Annexes I and II of the Habitat Directive (92/43/EEC) and Annex 1 of the Birds Directive (79/409/EEC) were recorded in the Central Training Area (Table 1).

Table 1. List of rare species, communities and habitats recorded in the Central Training Area, 2001.

PLANTS

Lithuanian Red Data Book, 2000 **I (E) category** – endangered species			
1	Rattlesnake Fern	*Botrychium virginianum* (L.) Sw.	
2	Red Helleborine	*Cephalantera rubra* (L.) Rich.	
3	Siberian Dragon's Head	*Dracocephalum ruyschiana* L.	
Lithuanian Red Data Book (2000) **II (V) category** – vulnerable species			
4	Grass-leaf Sandwort	*Arenaria saxatilis* L.	
5	Fringed Knapweed	*Centaurea phrygia* L.	
6	Indian Shoe	*Cypripedium calceolus* L.	**Annex II HD**
7	Northern Coral-root	*Corallorhiza trifida* Chatel.	
8	Dark-red Lesser Helleborine	*Epipactis atrorubens* (Hoffm.) Besser	
9	Cross Gentian	*Gentiana cruciata* L.	
10	Fragrant Orchid	*Gymnadenia conopsea* (L.) R. Br.	
11	Single-leaved Bog Orchid	*Malaxis monophyllos* (L.) Sw.	
12	Charity Greek Valerian	*Palemonium caeruleum* L.	
13	Large Self-heal	*Prunella grandiflora* (L.) Scholler	
14	Bractless Bastard Toadflax	*Thesium ebracteatum* Hayne	**Annex II HD**
Lithuanian Red Data Book (2000) **III (R) category** – rare species			
15	Leather Grape Fern	*Botrychium multifidum* (S. G. Gmel.) Rupr.	
16	Bulbiferous Coralwort	*Cardamine bulbifera* (L.) Crantz	
17	Thistle Broomrape	*Orobachne reticulata* Wallr.	
18	Narrow-leaved Lungwort	*Pulmonaria angustifolia* L.	
19	Small Scambiosa	*Scabiosa columbaria* L.	
20	Handsome Woollywort	*Trichocolea tomentella* (Ehrh.) Dumort.	

PLANTS

Lithuanian Red Data Book (2000) **IV (I) category** – rare, insufficiently known species			
21	Spotted Palmate Orchid	*Dactylorhiza ▯ entile* (Druce) Soó	
Lithuanian Red Data Book (2000) **V (Rs) category** – restored species			
22	Bears Onion	*Alium ursinum* L.	
23	Fir Clubmoss	*Huperzia selago* (L.) Bernh. Ex Schrank et Mart.	
24	Water Purslane	*Peplis portula* L.	
25	Polypetalous Meadow-grass	*Poa remota* Forselles	
26	Eastern Pasque-flower	*Pulsatilla patens* (L.) Mill.	**Annex II HD**
Habitat Directive **Annex II**			
27	Sand Pink	*Dianthus arenarius* L.	

PLANT COMMUNITIES

Lithuanian Red Data Book of Plant Communities (2000) **III I category** – rare communities			
1		*Ranunculo trichophylli-Sietum ▯ entiles* Th. Müller	

INSECTS

Lithuanian Red Data Book (2000) **III I category** – rare species			
1	Dingy Skipper	*Erynnis tages* L.	
2	Marsh Fritillary	*Euphydryas aurinaria* Rott.	**Annex II HD**
3	Green-underside Blue	*Glaucopsyche alexis* Poda	
4	Tree Grayling	*Hipparchia statilinus* Hfn.	
5	Large Blue	*Maculinea arion* L.	
6	Nickerl's Fritillary	*Melitaea ▯ entile* Nick.	
7	Turquoise Blue	*Polyommatus dorylas* D. & S.	
8	Olive Skipper	*Pyrgus serratulae* Ramb.	
Lithuanian Red Data Book (2000) **IV (I) category** – rare, insufficiently known species			
9	Scarce Heath	*Coenonympha hero* L.	
10	Swallowtail	*Papilio machaon* L.	

AMPHIBIANS

Lithuanian Red Data Book (2000) **IV (I) category** – rare, insufficiently known species			
1	Fire-bellied Toad	*Bombina bombina* L.	**Annex II HD**
2	Warty Newt	*Trutirus cristatus* Laur.	**Annex II HD**
Lithuanian Red Data Book (2000) **V (Rs) category** – restored species			
3	Natter Jack	*Bufo calaminta* Laur.	

BIRDS

Lithuanian Red Data Book (2000) **I (E) category** – endangered species			
1	Black-throated Diver	*Gavia arctica* L.	**Annex I BD**
Lithuanian Red Data Book (2000) **II (V) category** – vulnerable			
2	Boreal Owl	*Aegolius funereus* L.	**Annex I BD**
3	Stock Dove	*Columba oenas* L.	
4	White-tailed Eagle	*Heliaetus albicilla* L.	**Annex I BD**
5	Black Kite	*Milvus migrans* Bodd.	**Annex I BD**
6	Osprey	*Pandion heliaetus* L.	**Annex I BD**
7	Three-toed Woodpecker	*Picoides tridactylus* L.	**Annex I BD**
8	Grey-headed Woodpecker	*Picus canus* Gmel.	**Annex I BD**
9	Wood Grouse	*Tetrao urogallus* L.	**Annex I BD**
10	Hoopoe	*Upupa epops* L.	
Lithuanian Red Data Book (2000) **III I category** – rare species			
11	Shoveler	*Anas clypeata* L.	
12	Tawny Pipit	*Anthus campestris* L.	**Annex I BD**
13	Black Stork	*Ciconia nigra* L.	**Annex I BD**
14	Hobby	*Falco subbuteo* L.	
15	Crane	*Grus grus* L.	**Annex I BD**
16	Great Grey Shrike	*Lanius excubitor* L.	
17	Black Grouse	*Lyrurus tetrix* L.	**Annex I BD**
18	Honey Buzzard	*Pernis apivorus* L.	**Annex I BD**
19	Green Woodpecker	*Picus viridis* L.	
Lithuanian Red Data Book (2000) **IV (I) category** – rare, insufficiently known species			
20	Goshawk	*Accipiter* [illegible] *entiles* L.	
Birds Directive **Annex I**			
21	Lesser Spotted Eagle	*Acquila pomarina* C. L. Brehm	
22	Hazel Grouse	*Bonasia bonasia* L.	
23	European Nightjar	*Caprimulgus europeus* L.	
24	Red-breasted Flycatcher	*Ficedula parva* Bechst.	
25	Red-backed Shrike	*Lanius collurio* L.	
26	Wood Lark	*Lullula arborea* L.	
27	Slavonian Grebe	*Podiceps auritus* L.	
28	Barred Warbler	*Sylvia nisoria* Bechst.	

Table 1. (Continuation) List of rare species, communities and habitats recorded in the Central Training Area, 2001.

MAMMALS

Lithuanian Red Data Book (2000) **III I category** – rare species			
1	Pond Bat	*Myotis dasycneme* Boie	**Annex II HD**
Lithuanian Red Data Book (2000) **IV (I) category** – rare, insufficiently known species			
2	Serotine Bat	*Eptesicus serotinus* Schreber	
3	Mountain Hare	*Lepus timidus* L.	
4	Great Noctule	*Nyctalus noctula* Schreber	
Lithuanian Red Data Book (2000) **V (Rs) category** – restored species			
5	Otter	*Lutra lutra* L.	**Annex II HD**
Habitat Directive **Annex II**			
6	European Beaver	*Castor fiber* L.	

HABITATS

Habitat Directive **Annex I**		
1	2330	Inland dunes with open grasslands
2	3160	Natural dystrophic lakes
3	3260	Water courses with mole-cricket communities
4	4030	Heather communities
5	7140	Transitional bogs and quaking bogs
6	7160	Mineral-rich springs and springfens
7	9010	Western taiga
8	9050	Herb-rich forests
9	91D0	Bog woodland
10	91E0	Alluvial forests

Development of nature management plan. The inventory data were analysed and 10 sites of great natural importance were selected in the territory of the Central Training Area (Table 2). Criteria for identification of the mentioned sites agree with those usually used by scientific experts while preparing nature management plans in civil territories:

- site is recommend as a NATURA 2000 site,
- abundant populations of species included in the Habitat Directive (Annex II) and the Birds Directive (Annex I) breed or are regularly recorded,
- important areas of habitats included in the Habitat Directive (Annex I) are recorded,
- abundant populations of species that belong to categories 0, 1, 2 and 3 of the Lithuanian Red Data Book are recorded,
- important areas of communities that belong to categories 1 and 2 of the Lithuanian Red Data Book are recorded.

Table 2. Sites of great natural importance and their conservation measures in the Central Training Area.

	Sites of great natural importance	Conservation measures
1) MAINTENANCE OF CURRENT NATURAL CONDITIONS		
1	**Lake Varnakėlis and bog woodland** Valuable elements: 3160 Natural dystrophic lakes 7140 Transitional bogs and quaking bogs 9010 West taiga 91D0 Bog woodland	- preserve current water regime - preserve current forest state - limit walking - construct one wooden walk towards the lake instead of present stone path
2	**Dry pine forest** Valuable elements: 9010 West taiga	- preserve integrity of forest stand - ban on driving - preserve dead and fallen trees
3	**Chain of Lakes Liūnukas and bog woodland** Valuable elements: 91D0 Bog woodland rare butterfly species typical of a raised bog rare birds: black-throated diver, slavonian grebe, wood grouse, crane, great grey shrike	- preserve current water regime - limit walking - restricted visit during bird-breeding period (March-May) - ban on intensive forestry
4	**Valley of the Bevara rivulet** Valuable elements: 3260 Rapids of rivers with mole-cricket communities 7160 Mineral-rich springs and spring fens 9010 West taiga 9050 Fir forests rich in herbage 91E0 Alluvial forests rare plant: polypetalous meadow-grass rare amphibian: warty newt rare birds: grouse, crane, boreal owl, stock dove, three-toed woodpecker, red-breasted flycatcher, black stork, goshawk, lesser spotted eagle, grey-headed woodpecker rare animals: otter, white rabbit, beaver	- reduce number of beavers (one family per rivulet) to preserve plant communities characteristic to fast-running rivulets and communities of springs - preserve dead and fallen trees – the wintering sites for amphibians and nesting sites for birds - preserve current water regime - ban on forest cutting - ban on forest activities during bird-breeding period
5	**Lake Molinė and surrounding forests** Valuable elements: 7160 Non-calcareous streams and swamps with streams rare plants: 23 species enlisted in the Lithuanian Red Data Book, 4 species enlisted in the Annex II HD: Indian shoe, bractless bastard toadflax, eastern pasque-flower, sand pink rare amphibian: warty newt rare birds: pern, black-cocks, grouse, crane, boreal owl, stock dove, grey-headed woodpecker, three-toed woodpecker, red-breasted flycatcher, goshawk, black stork	- apply nature reserve regime - eliminate coniferous trees and shrubs in the forest plots No. 88 and No. 108 periodically to preserve well illuminated habitats for some plants - ban on driving (even on roads and tracks leading towards Lake Molinė) - ban on forest cutting on slopes to reduce slope erosion - preserve dead and fallen trees – the wintering sites for amphibians - restricted visit during bird-breeding period (April-June) -ban on clear tree cutting and limited selecting cutting in the entire site

	Sites of great natural importance	Conservation measures
1) MAINTENANCE OF CURRENT NATURAL CONDITIONS		
6	**Fishery ponds Pažeimenė** Valuable elements: rare amphibians: warty newt, natter jack, fire-bellied toad rare birds: shoveler, white-tailed eagle, osprey, black kite	- ban on forest cutting - preserve dead and fallen trees – the wintering sites for amphibians - preserve current water regime and keep and the water at least in the 50 % of ponds by maintaining fish farming in the ponds - keep drainage ditches in case the fish farming will be stopped in the ponds area
2) MAINTENANCE OF OPEN AREAS		
7	**Shooting field** Valuable elements: 2330 Inland dunes with open grasslands 4030 Heaths rare plants: Eastern pasque-flower, grass-leaf sandwort, sand pink highly rare butterfly species typical for sandy areas rare birds: a large population of tawny pipits, wood lark, european nightjar, black grouse, hoopoe	- constant destruction of the integrity of plant cover to prevent formation of pine stands - remove all pine trees within 3-4 years - divide territory into 8-10 areas and apply controlled fires by rotation in late autumn (November) or late spring (February-March), before snowfall or after snow melt - stop forest activities during the bird-breeding period
8	**Tankodrom** Valuable elements: 2330 Inland dunes with open grasslands 4030 Heaths rare plants: eastern pasque-flower, grass-leaf sandwort rare butterflies: olive skipper, the sole population of the tree grayling in Lithuania	- maintain extensive military training - remove majority of pine trees
9	**Autodrom** Valuable elements: rare butterflies: swallowtail, large	- prevent overgrowing with pine trees - mow the glades
10	**Territory around the entrance-exit control point** Valuable elements: rare butterflies: olive skipper, marsh fritillary	- prevent overgrowing process - restore and maintain glades

Conservation measures for protection of valuable botanical and zoological objects are based on the peculiarities of each species, community or habitat type. In addition, natural as well as anthropogenic factors, causing unfavourable survival conditions for rare species, were identified. The main natural problem was degradation of heath lands and open inland dunes due to overgrowth processes. In order to undertake proper conservation measures the sites of great natural importance were assessed in regard to the influence of military activities and area utilisation by considering whether military utilisation is damaging, insignificant or beneficial (or even necessary) to a natural system.

Two main types of conservation measures for the sites of great natural importance were discerned in the Central Training Area (Table 2, Figure 1):

1) maintenance of current natural conditions through activity restrictions, avoidance of any changes of existing natural conditions and implementation of special conservation measures,

2) maintenance of open areas through control of overgrowth processes by cutting out new tree saplings or creating local burn-outs and controlled disturbance of plant cover on periodical base as well as by mowing meadows.

The first type of conservation and maintenance measures were planned for those sites of great natural importance, which military impact is influencing or might influence negatively and for those, to which military utilisation is insignificant just in order to ensure prevention of damages to natural systems. Military training and access must be avoided in these sites covering rare forest and bog biotopes, also lakes and rivulets (sites of great natural importance No. 1-6).

The second type of conservation and maintenance measures were planned for those sites of great natural importance, to which maintenance of military training and economic activities are beneficial and even necessary. Military training must be maintained in the sites covering rare dry heath land and open inland dune biotopes (sites of great natural importance No. 7-9) and economic activities must be maintained in valuable meadow biotope (site of great natural importance No. 10).

A complex coordinated approach that focused on the entire site of great natural importance was used while planning conservation and maintenance measures in the Central Training Area, rather than an approach focused on the preservation of single species. This complex approach enables the maintenance of habitats and conditions favorable for many rare plants and animals at one time. For instance, periodical disturbance of ground surface and plant cover in the shooting field of the Central Training Area assists in maintenance and preservation of such rare habitats as heath lands and open inland dunes, which in turn provide such rare species as Eastern Pasque-flower, Sand Pink, Tree Grayling, Olive Skipper and Tawny Pipit with necessary favourable conditions for their existence and proliferation.

Implementation. Implementation of the nature management plan of the Central Training Area is foreseen for the period of 2003-2007. Conservation measures were graded according to priorities from the first to the fifth year in an action plan. Most of the conservation measures related to restrictions, constant limited or extensive intervention will not end in the certain year but will last for the entireimplementation period and will be prolonged in the revised plan later on if necessary.

Monitoring and revision of the nature management plan. Monitoring and assessment of the populations of rare animal and plant species and habitats will be conducted every 3-5 years in compliance with the National Monitoring Program for the Sites Important for Conservation of Habitats and Birds (2002). Despite the fact that the first rare species and habitats monitoring will be conducted in 2008, inspections on abundance of some sensitive protected species after the implementation of the special protection measures (restoration of open areas by cutting pine tress) within their habitats will be carried out annually.

In addition, monitoring of effectiveness of the conservation measures will be conducted in order to revise the plan and improve management. Two aspects will be evaluated: the quality of execution of a conservation measure (++ excellent, + good, 0 neutral, - satisfactory, -- unsatisfactory) and the relevance of a conservation measure to habitats and species (++ very relevant, + relevant, - inadequate). It is very important to consider both aspects, because sometimes it might happen that execution quality of a measure is excellent, but the impact of the measure is negative. Proper evaluation will ensure the right corrective actions and update of the nature management plan.

Forest areas were not considered in this plan, as these areas were described in terms of management and forestry activities separately in the Forest Management Plan of the Central Training Area.

CONCLUSIONS

The pilot project in the Central Training Area as well as experience from military territories of other countries shows that military training areas maintain a great biological diversity primarily because of specific military training activities, relative inaccessibility by the public as well as because fertilizers and pesticides are not used. Thus military utilisation can be maintained in order to preserve rare natural objects unique to military training areas and rare in the countries of Europe. But the continued presence of rare fauna and flora in military training areas requires very well balanced and careful use of military land with proper conservation measures and regular observations and evaluations.

The vital moment in nature management planning is continuity of this process. One-time management "injections" do not principally improve conditions for rare species and habitats.

Slovak Military Veterinary Services Protect and Exploit Rare Species for 11 Years

Juraj Marcin, and Jozef Németh
Military Veterinary Institute, Košice, Slovakia

INTRODUCTION

In the time when man first realized his abuse of nature the question where is border between animal exploitation and protection of natural life emerged.

We, together with animals occupy the same biotop and we are only a tiny part of this world. Our ability to use the brain differently from other living creatures makes us responsible for protection and rational exploitation of wildlife.

Slovak Armed Forces exploit rare falcons, especially sakers (Falco cherrug) and peregrines (Falco peregrinus). This wonderful creatures are exploited in Air Force bases by falconers who prevent bird strikes with aircrafts.

Military Veterinary Institute in Košice is the authority in Slovak Armed Forces solving this problem.

Military Veterinary Institute deals with following tasks:

1. Food safety inspections, research in the field of foodstuff.
2. Epidemiology and rapid diagnostic methods of pathologic agents.
3. Ecology, sanitation and bird strike hazard assessment.

Ecology, Sanitation and Bird Strike Hazard Assessment Department is responsible for all environmental issues concerning military activities in military bases or training areas, predominantly in the area of biological contamination

The Military Veterinary Institute is military methodological and statistical authority for bird strikes.

K. Mahutova et al. (eds.), Defense and the Environment: Effective Scientific Communication, 217–221.

Ecology and bird strike hazard assessment represents two sides of the same problem.

Despite the events that happened on September 11, 2001, and in spite of present airline crisis, civil air traffic takes great deal of cargo and fare transportation, especially long distance transportation. The situation in military aviation worldwide is quite different because of considerable increase of hours flown by military aircraft. Along with, there are strong demands for top air traffic safety and a reduction in flying incidents and fatal crashes.

Animals, especially birds, don't recognize fences, borders or manmade restrictions. Slovak Air Forces released information that bird and animal strikes with aircraft is the second most frequent reason for flying incidents. International Civil Aviation Organisation (ICAO) published information that most of bird strikes happened in Europe (42 %) and in North America (32%). Another ICAO statistics show that 79% of bird strike incidents happened at airports, 11% in vicinity of airports, and only 10 % outside airport areas. Bird strike statistics of our Institute suggests that the most endangered parts of aircraft are the engines (33%), the canopy (29%), and the wings (10%). Bird strikes occurred mostly on approach, although a significant number happened during take-off run, climb-out, and landing roll. The greatest number of bird strikes happened between ground level and 100 feet above the ground, when the aircraft is at an indicated airspeed of between 100 and 150 knots. The collision more often happens with a single bird than with the flock of birds.

ICAO sources have reported over 95,000 bird strike events since 1995 and over 130 casualties were confirmed. We can see from the statistics that bird strikes are very significant cause for fatal and costly accidents. For example, in 1995, after a bird strike at JFK airport in New York, the repair price for Concorde was US $9 million and compensation was US$5.3 million, totalling US$14.3 million. Or, in August 2000, near Los Angeles airport, the amount of petrol dumped from B-747 into ocean due to bird strike was shocking 83,000 tons.

And some more significant from the military point of view, in 1996, a NATO E-3 AWACS crashed in Greece. This was non-harmful bird strike accident, as for no casualties were reported. But in 1995, a similar bird strike accident in Alaska resulted in 24 casualties on board.

The exploitation of rare birds of prey in airfields can reduce number of birds. This is a main task for Bird and Animal Control Units at airports.

Special instruction for wildlife hazard assessment was published for civilian traffic in ICAO paper Annex 14 volume 1. For military use similar procedure were released in NATO Instruction Paper STANAG 3879 FS.

Main missions Bird Control Units should provide are:

1. Detecting animals and birds.
2. Repelling animals and birds.
3. Implementing preventive measures at airfields.
4. Providing information to pilots and flight control.

The most important part is repelling animals, especially birds. One of effective methods is falconry. Sure, a much simpler method would be hunting and killing everything flying near the airport and runways. But in our country over 85 % birds occurring in airfields are protected by law. Therefore, it appears that falconry presents effective and ecological method in bird strike prevention.

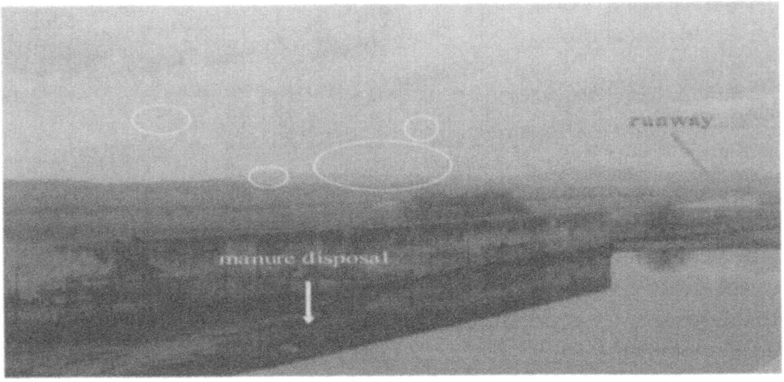

This photograph is typical example of the close relation between air traffic and ecology in airport vicinity. Manure disposal is very close to the runway and flock of birds in the altitude of about 60-80 meters, an altitude where aircraft would fly through after take off. Why do birds occur here? It is so simple. Manure presents food source. It is an ideal place for evolution of many insect species and worms, larvae, and matured insect are bird food source. And increased number of birds is a risky factor for aircrafts, thus creating the need for bird and wildlife hazard control units at Air Bases.

Falcons are natural predators, effective to control certain kinds of birds and mammals and falconry is also impressive to public. But international law protects these animals since they are an endangered species.

Due to these problems, in 1991, Slovak Armed Forces, in cooperation with the Veterinary University in Košice, have built and equipped special Falcon Breeding Centre (FBC) in Rozhanovce. This place was carefully chosen in wooded country to be free from civilization disturbances. In the winter of 1991, 5 breeding pairs of falcons were imported from the Czech Republic. They were used to start the project. Since the start of the project until present days, the three man staff has been securing the facility. They were skilled in keeping birds of prey, but they had to learn, in extremely short period of time, the wide number of activities necessary for a successful breeding program. There are several critical points where breeding can fail. Success depends on:

- choosing the right individuals for mating
- preparing for laying
- laying and incubating eggs
- hatching
- releasing young falcons in new sites

In 1992 we have started project partially supported by Ministry of Environment called "Falcon Reintroduction in Wilderness". In 1994, artificial insemination was launched to increase the efficiency of breeding.

In 1995, (FBC) became a rehabilitation center for injured and handicapped animals, a project supported financially by Slovak Ministry of Environment.

We tried storing sperm frozen, in a liquid nitrogen container, but it wasn't very successful because of lower sperm vitality. We began using sperm immediately after taking. It is necessary to check the sperm for vitality and cleanness. Female falcon is fertilised with the micropipette. Artificial insemination enables to get approximately 4-5 fertilized eggs. These are taken from the bird's nests and put into an incubator. After 30-32 days, at 38 degrees Celsius and moisture of 74%, new life is born.

After few days, in incubator new individuals are put into warmed nests and here they are bred up to the age of 6 weeks, either alone or together with foster parents, when we have to decide whether it will be released into wilderness or will serve for military purposes. Youngsters to be released into the wild must not be dependant on human in their age of imprinting. If returned to nature, they are given to natural nests in wilderness.

We have a project of reintroduction Falco Peregrinus, the peregrine, into municipal communities where it is necessary to observe whether they are able to hunt and survive. In the city, peregrines are put into wooden cages in non-disturbing area where there are with rich food opportunities available, like pigeons.

This work is done in cooperation with State Environmental Inspections as the supervisors and state authorities.

SUMMARY

In 10 seasons we have produced 122 individuals. Eighty (80) of them were returned to nature: 44 sakers (Falco cherug) and 36 peregrines (Falco peregrinus). Forty-two (42) birds of prey were deployed in Slovak Air Force Bases. At present, we have 31 breeding birds in FBC.

Project of reintroduction of these rare birds of prey into Slovak environment was successful and tax payers money was returned to the people through enriched nature. Furthermore travellers by plane surely appreciate services provided by these beautiful creatures at Slovak airports where they improve flight security.

This co-operation between Ministry of Defence and Ministry of Environment is the only right way of possible communication with the public.

REFERENCES

1. ICAO: Annex 14 Volume I Aerodrome Design and Operations Montreal 1990
2. ICAO: Airport Services Manual ,part 3-Bird control and reduction, ICAO Doc 9137-An/898, Montreal 1991
3. ICAO: Manual on the ICAO Bird Strike Information System(IBIS),ICAO Doc9332-An/909,Montreal 1989

Protection of Natural Habitats within the Military Land, Experiences from U.S. Army, Europe

Wolfgang Grimm
U.S. Army Installation Management Agency, Europe Region, Engineer Division, Environmental Branch, Heidelberg, Germany

INTRODUCTION

The United States Army in Europe (USAREUR) has implemented a comprehensive and effective environmental program with the overall goal to integrate environmental stewardship into all military activities. On military lands, natural habitats must be managed and protected while simultaneously supporting military mission requirements. USAREUR's pro-active natural resources management and the implementation of the Integrated Training Area Management concept proved that environmental protection and military land use are not mutually exclusive but, on the contrary, of a symbiotic relationship. The European Union's "Flora, Fauna, Habitat Directive" is one of the new challenges facing the military in Europe. Army natural resources managers are challenged to keep the Army in legal compliance while simultaneously preventing or minimizing negative impacts of environmental legislation on the effective conduct of its military training mission.

USAREUR's environmental stewardship is built on four pillars: Compliance, Prevention, Restoration, and Conservation.

The objective of the compliance program pillar is to attain and sustain compliance with all host nation and U.S. requirements; Restoration is to remediate contaminated sites as effectively as resources permit to protect human health and the environment; Prevention adopts and implements management approaches, procedures, and operations to minimize all environmental contamination and pollution; and finally, Conservation is to conserve, protect, and enhance natural and cultural resources.

The overall goal of the U.S. Army's environmental program is to integrate environmental stewardship into all military activities.

THE LEGAL BASIS

U.S. Army Europe operations are governed by the NATO Status of Forces Agreement (SOFA) which is the international agreement designed

K. Mahutova et al. (eds.), Defense and the Environment: Effective Scientific Communication, 223–230.

to serve mutual interests of both, the Army and the Host Nations. It outlines relationships on matters from criminal jurisdiction to customs, taxes and other civil and legal matters. The most recent German Supplementary Agreement to the NATO SOFA is significant because it dictates German law is applicable on our installations. This along with EU regulations and host nation-implemented EU directives (because they are considered valid Host Nation (HN) requirements), and applicable US law and Department of Defense policy - Final Governing Standards (FGS) - make up the body of environmental program guidance from which we derive our requirements.

The Final Governing Standards are a comprehensive set of country-specific, substantive, environmental provisions. They are developed by comparing and adopting the more protective requirements of the Overseas Environmental Baseline Guidance Document "OEBGD" and HN environmental standards, and applicable international agreements. They consist of 18 environmental protocols:

1	Environmental. Progr. Mgmt.	10	Environmental Noise
2	Air Emissions	11	Pesticides
3	Drinking Water	12	Cultural Resources
4	Wastewater	13	Natural Resources
5	Hazardous Materials	14	PCBs
6	Hazardous Waste	15	Asbestos
7	Solid Waste	16	Underground Storage Tanks
8	Medical Waste	17	Lead Based Paint
9	Petroleum, Oil and Lubricants	18	Spill Prevention & Response

One of the FGS key compliance requirements for protocol # 13, Natural Resources, is that natural habitats of indigenous species must be protected. For that purpose the U.S. Army in Europe has established and implemented integrated natural resources management plans for all its installations. These plans contain specific provisions and guidelines for the protection of natural habitats on military lands. The plans are called "integrated" because they are closely coordinated with military trainers to assure they are in support of and not in conflict with the military training mission. The plans consist of three volumes: (1) general information about the installation, (2) natural resources protection in cantonment areas, and (3) management and protection of natural resources, flora and fauna species and their habitats on military training areas and other mission lands. The integrated natural resources management plans will also serve as the basis for the Flora, Fauna, Habitat (FFH) Management plans, which have to be implemented in accordance with the European Council Directive 92/43. Since this EU directive, which was incorporated into German law in 1998, is a new challenge the military in Europe is faced with, this subject will be discussed in greater detail.

THE FFH CHALLENGE

In 1992, the European Union decided to establish a coherent European ecological network under the title "Natura 2000", which is composed of sites as required by the European Union Council directive of May 1992 (Flora, Fauna, Habitat) and the European Union Council Directive on Bird Protection. The aim of the directive is to contribute toward a guarantee of bio-diversity by conserving the natural habitats of fauna and flora species. Each European Union member state was required to report its areas to be placed under protection. The site selection was based strictly on scientific criteria. The current site use was not considered. Therefore, military training areas, or other areas used for defense purposes, were not excluded. It is up to each member state to establish measures to protect natural habitats and species at reported sites. However, interpretations on whether the protection measures are sufficient to meet the directive requirements are subject to the European Court of Justice. The directive was adopted without exceptions or allowances for the requirements of military training. Therefore, we are concerned that designation of military areas under the directive may have a significant impact on the use of these properties. While the present use of a nominated area may continue unrestricted if the overall ecological value is not degraded, we are concerned that future modifications, like the construction of a new, or the extension of an existing firing range might be significantly delayed, or in the worst case, disapproved. We will use the integrated natural resources management plans to manage and conserve resources to the maximum extent possible. On one hand, our position is to maintain good stewardship of our training area environments. On the other hand, we must retain the flexibility to adjust to changing training requirements.

In 1998, the EU directive was incorporated into German law. Based on this law, the German States were tasked to nominate areas to the German government, who reported these sites to the EU for inclusion into the "Natura 2000" network. In order to safeguard military interests, the German government added a reservation that nomination must not result in significant impacts on the continued military use, including potential usage changes. However, we are not sure whether the German government reservation will pass scrutiny by the European Court of Justice. The military reservation reads: "The areas mentioned are subject to military use. Through this nomination and a later inclusion of these areas in the list of sites of community importance no essential restrictions of the continued use, including a change in the use of these areas for the purpose of defense of the Alliance or national defense, may occur". The

European Commission's opinion is that the directive does not provide a basis for excluding areas used for defense purposes.

Under applicable German law, the States (Länder) are ultimately responsible for implementation of the EU Directive 92/43, including the establishment of FFH management plans. On federally-owned land, the vast majority of lands used by the U.S. Army, the Länder will enter a formal agreement with the Federal Republic of Germany delegating the establishment of FFH management plans to the federal forest service. In order to safeguard US military interests and to be in legal compliance, the U.S. Army Installation Management Agency, Europe Region and the German Federal Ministry of Finance agreed on the following procedure: The US and the federal forest service will jointly prepare the plan. The US commander will approve the plan. The federal forest service will forward the US-approved plan as the official FFH management plan under their cover memorandum to the State (Land). On non-federal land, the states will prepare the management plans. Through participation in the development process we must ensure that military interests are covered. Due to its proactive conservation program, the U.S. Army Installation Management Agency, Europe Region (IMA-E) has a wealth of threatened and endangered species data available, which will be the basis for future FFH management plans. Since military land use in combination with sound natural resources management is the reason for the high ecological value of training lands, we are questioning the necessity for an additional FFH protection layer. However, we will continue our cooperation with the host nation governments and act as a good steward of the lands entrusted to us. Our main objective is to protect natural resources while retaining the flexibility to maintain military readiness.

SUPPORT OF MILITARY TRAINERS WHILE PROTECTING NATURAL HABITATS

The most important support which environmentalists can provide to the trainers is to keep them in environmental compliance. Environmentalists must keep the trainers out of legal trouble, so that they can fully concentrate on their mission, which is to conduct military training. Another very important environmental support function is to provide training land for realistic and effective military training on a sustainable basis. Training land is the "classroom" for soldiers in which they are taught how to fight, survive, and win a battle. Training lands have become scarce and additional training space is not available. We must keep training lands available for optimal use not only for us, but also for future generations. This principle of sustained training land availability includes two environmental objectives: (1) protect natural habitats as legally required and (2) maintain healthy ecosystems as the prerequisite for realistic and natural training scenarios.

INTEGRATED TRAINING AREA MANAGEMENT AND COOPERATION WITH TRAINERS

The implementation of the integrated training area management (ITAM) program is the initial and most important step for environmental planning and conducting training. It assures that environmental protection supports the training mission rather than hindering it. Trainers and environmentalists must have the following common goals: support the Army training mission, sustain training lands, manage and protect natural resources.

In 1985, ITAM was established at Hohenfels major training area, which was a critical step to integrate protection of natural resources into training mission requirements. Today, this program is implemented at all U.S. Army training areas in Europe. It builds the bridge between the training and the environmental community. ITAM is a team effort: The military trainers (G3) have the lead, while environmental personnel provide technical expertise. ITAM consists of four program elements: Land condition trend analysis (LCTA), land rehabilitation and maintenance (LRAM), training requirements integration (TRI) and sustainable range awareness (SRA).

LCTA is a management procedure that inventories and monitors land conditions. It incorporates relational database and Geographic Information System (GIS) technologies into the land-use decision process. LCTA collects physical and biological resources data from training land in order to relate land conditions to training and testing activities. These data provide the information to effectively manage land use and natural and cultural resources.

LRAM is a preventive and corrective land rehabilitation and maintenance procedure that reduces the long-term impacts of training and testing on an installation. It mitigates training and testing effects by combining preventive and corrective land rehabilitation, repair, and/or maintenance practices. It includes training area redesign and/or reconfiguration to meet training requirements.

TRI is a decision support procedure that integrates all requirements for land use with natural and cultural resources management processes. TRI integrates the installation training and testing requirements for land use derived from the Range and Training Land Program (RTLP); the range operations and training land management processes; and the installation training readiness requirements with the installation's natural resources conditions. The Army Training and Testing Area Carrying Capacity (ATTACC) program is the standard ITAM methodology for estimating training land carrying capacity by relating training load, land condition, and land maintenance practices. The integration of all requirements occurs through continuous consultation among the Directorate of Plans, Training, and Mobilization (DPTM), natural and cultural resources managers, and other environmental staff members. The output of the TRI process is incorporated in the installation's Integrated Natural Resources Management Plan (INRMP).

SRA provides a means to educate land users on their environmental stewardship responsibilities. It provides for the development and distribution of educational materials to land users. These materials relate the principles of land stewardship and the practices of reducing training and/or testing impacts. SRA also includes information provided to environmental professionals concerning operational requirements.

ITAM applies modern technologies, such as remote sensing, geographic information systems, global positioning systems, land condition trend analysis, land carrying capacity evaluation, erosion modeling, helicopter seeding, irrigation test stations and surface water measuring systems.

ENVIRONMENTAL PLANNING

Environmental managers must be familiar with the requirements of a successful military training exercise. Without this knowledge, environmentalists cannot optimally support the trainers. Trainers must identify their specific requirements to enable the environmentalist to provide an optimal “classroom”, as training lands are called. On the other hand, environmentalists need a sound knowledge of natural resources features on the training area. For that purpose, environmental baseline surveys are performed focusing on threatened and endangered species, fauna, flora, soil, topography, vegetation communities, surface water, hydrology, and wetlands. Based on the survey results, and after having identified the training area carrying capacity (i.e. what pressure can an area take without being irreversibly damaged?), it is possible to assign the best suited land parcels to individual training missions while protecting natural habitats to the maximum extent possible.

THREATENED AND ENDANGERED SPECIES MANAGEMENT

The United States Army, Europe has implemented a comprehensive and effective environmental program beginning in the late seventies. One of the fundamental program elements is the management of natural resources, including the protection of threatened flora and fauna species. As a result of military land use, combined with successful natural resources protection efforts, our training lands have developed into areas with high ecological value, which made them prime candidates for nomination under the European FFH. Training areas have become very important retreat areas for threatened and endangered species. They are ecological islands surrounded by built-up areas and intensively used farm and forest lands. In U.S. military training areas, very limited amounts of fertilizers and no pesticides are applied, thus increasing the abundance of flora and fauna species. Another important ecological factor is that all stages of natural succession, from bare ground to old-growth forests, are present on training areas. Training areas, therefore, provide homes for more flora and fauna species than hardly any other land in the densely populated and managed areas of central Europe. Military training and environmental protection are not mutually exclusive. On the contrary, the enormous amount of threatened flora and fauna species thriving on military lands documents a symbiotic relationship.

The U.S. Army's mandatory integrated natural resources management plans require threatened and endangered species surveys be conducted for all training areas and other mission land. The results of the species surveys were overwhelming. They revealed an incredibly high number of species, listed in the federal and state red list data books, thriving on military lands. Each survey includes management recommendations for the species found on the installation. These management recommendations ensure that through the ITAM program's Training Requirements Integration element, species are protected to the maximum extent that effective military training allows. Photo documentation allows non-biologists to identify species found on the installation. Generally the following species groups are surveyed: Plants, Amphibians, Reptiles, Ants, Bats, Birds, Bugs, Butterflies, Dragonflies, Bees, Locusts, and Moths. The location of species found are shown on a digital map, which can be overlaid on a GIS system to other map layers, such as topography, soil erodability, surface waters, installation boundaries, etc. Clicking on a species location shown on the map opens a database with detailed species information and management recommendations. This GIS capability produces maps for training scenario writers, identifying areas for conduct of doctrine-compliant training with minimal impacts on the environment.

OUTREACH PROGRAM

To ensure unrestricted conduct of military training, the existence of training areas should be accepted by the civilian neighbor communities. It has shown to be beneficial to play with open cards and not to hide behind installation boundaries. The former Commanding General, U.S. Army, Europe, has conducted two environmental orientations at the two major training areas in Grafenwoehr and Hohenfels. He invited politicians from the European Union level, German federal and state governments and local mayors to attend briefings and a field trip through the training areas to demonstrate that the U.S. Army takes its environmental protection obligations seriously.

Today, USAREUR is more active in publicizing environmental success stories than in previous years. This has proven to be a smart move since its reputation as a good environmental steward has increased significantly. The Army even participates in public environmental awards programs. The 293d (Mannheim) Base Support Battalion was recognized with three environmental stewardship awards by the City of Mannheim. These public relations initiatives and the consequent integration of environmental protection efforts into mission requirements raised the acceptance level of the local public for the existence of our training installations.

COOPERATION WITH HOST NATION

Close cooperation with host nation authorities is essential for a successful military environmental program. This is true for all levels from federal ministries, state ministries, district governments, and counties down to the local community. Environmental topics of fundamental importance are discussed at least once per year at federal and state government level in environmental working groups. The federal forest service is a reliable partner for all matters of natural resources management. Forest service and Army personnel jointly participate in annual natural resources managers training seminars which the Army hosts. Personal contacts at all levels help to solve items of concern immediately, before they turn into problems.

SUMMARY

The U.S. Army in Europe has taken a proactive approach towards integrating environmental protection into all military activities. Military training areas have developed into lands of extremely high ecological value. They have become the most important retreat areas for threatened and endangered flora and fauna species. The abundance of species demonstrates that military land use and species protection are not mutually exclusive, but form a symbiotic relationship. Military training and the subsequent readiness of troops have highest priority. The integration of professional natural resources protection into military mission requirements assures continued existence of natural habitats while simultaneously providing classrooms for realistic and effective military training on a sustainable basis.

The Problem of the Environment and Nature Conservation in Romania; *Case Study on Shrublands as Nature Protected Areas and Military Camouflage*

Doina Alexandra Botzan[1], Mihaela Paucă – Comănescu[2]
[1]*National Research and Development Institute for Environmental Protection*
[2]*Institute of Biology, Romanian Academy*

INTRODUCTION

Environmental Protection is an important point of Romanian the integration program into the European Community. In this context it was accepted the community acquis about Chapter 22, Environmental Protection. The Romanian laws will be harmonized with EU legislation.

In Romania, the legislative process about nature conservation started in 1926. At that time Floodplain of Low Danube reservation was created. The South Carpathians -Retezat Mountains reservation was created in 1935. In the last ten years laws system was improved and national and natural parks are stipulated into law.

The ecosystem structure of shrublands corresponds to camouflage needs for certain military objectives but at the same time to the nature conservation status in mountain region.

ENVIRONMENTAL LEGISLATION

In the context of adhering to the EC, Romania accepts the community acquis about Chapter 22, Environmental Protection, come into the operation at December 31, 2000. Romania unilaterally assumed January 1, 2007 as a working hypothesis for finishing the preparation for EU adhering. Romania will implement the community acquis until the adhering data.

From institutional point of view in the environmental protection field the Government Decision NR. 17/ 2001was adopted about organization and functioning of the Water and Environmental Protection Ministry. In the new structure, a 4-state secretary helps the minister in his activity, which activate in the next field:

- environmental protection

K. Mahutova et al. (eds.), Defense and the Environment: Effective Scientific Communication, 231–244.

- water management
- European integration
- Parliament relation

The State Secretary for European Integration coordinates the legislative harmonization activity of national legislation with community environmental acquis for the elaboration, promoting and controlled accelerating of the new implemented legislation.

In the scope of development of the dialog between government structure and civilian society, and for showing the real role and activity of Water and Environmental Protection Ministry, a Public Relation Department was created.

Environmental activity is led by the Environmental Protection Law no. 137/1995 modified and republished in 2002. In what concern nature conservation there are specific laws, which regulate this activity like:

- 462/ 2001 concerning protected areas regime, natural habitat of flora and fauna conservation;
- 5/2000 about national territory management;
- 82/ 93 concerning Danube Delta Biosphere reservation.

The law no.462/201 stipulates the categories of protected natural areas and natural habitat categories of wild flora and fauna. In the same time the law no. 5/2000 - about national territory management - named 17 objectives like biosphere reservation, national and natural parks as well 827 protected natural areas. Also included are about every other environmental factor: soil, water, atmosphere, and noise. Those laws together with 137/2002 provide the framework for nature conservation. The territorial inspectors are responsible for the application, monitoring and control of laws application.

STUDY CASE

The large diversity of vegetation in Romania is divided between three main formations:

-Forest
-Grassland
-Shrubland

The first and the second categories are spread on the largest area in Romania; they are the surfaces most taken care of because these bring money to the state economy and to private owners' pockets. The third category, that of shrubland, is the most endangered first because it is spread on more reduced areas, then because nobody considers it due the small economic interest; people prefer to change the destination of this land either as grassland or as forest and even for other uses.

The general qualities of all shrubland areas are:

- represent a reservoir of a large biodiversity of flora, fauna and different ecosystemic structures (habitats);
- represent a potential green corridor for the movement of wild animals (mammals, birds, reptiles as well as thousands of invertebrate species);
- protection against soil erosion because shrublands are currently located in pioneer conditions of life – high mountain or river banks in the plain region, frequently and long time flooded;
- They use the worst soil conditions for a plant, such as water deficit or excess, low trophicity, low structured soil and extreme, "limit" temperature.
- they have a perennial existence, being able to live and cover the same soil for tens and hundreds of years;
- shrubs can grow there where other wood vegetation can't!
- shrubland improve forest restoration as a stage in natural succession of vegetation, when this is necessary to be done;

Shrublands are spread throughout Romania; as primary and permanently renewable installation in the high mountain region (about 1650 – 1850 m and higher) and along the rivers all over the country; the areas along the rivers are of course larger in the plain region than in the mountain or hilly regions because floodplain land is larger there (Vegetation map). The Danube Delta and Black Sea shore are colonised by shrubs too.

Shrublands don't all belong to the same species. They are rarely monodominant and their associations are very diverse, consisting of different shrubs; sometimes-even dwarf trees are present and they are always accompanied by many herbaceous species.

The main associations of shrubland recognised by phytosociologists in Romania are presented in Table 1. Fourteen of them are primary installed, in a climax relation with the abiotic environment, the other 5 are secondary.

Some of the shrublands probably with the largest stretch are the mountain pine shrubs, in the subalpine belt of mountains. The next association as extension and as neighbourhood is juniperus scrub and it is the strip between subalpine and boreal forest belts, which groups pine shrubs and dwarf trees living there at the physiological limit of tree existence. In the same belt area there is a mountain juniper scrub, a little smaller, more rare than mountain pine, and preferring acid rocks instead the calcareous.

Table 1 Shrubland types and their distribution in Romania

No	SHRUBLAND TYPE	Vegetation zone and belts	Vege-tation type	Vegeta-tion map unit.	Widespread	Presence in protected area
PRIMARY VEGETATION						
1	***Rhododendro myrtifolii – Pinetum mugi*** Borza 53 em.Coldea 84 (Carpathian mountain pine shrub)	Subalpine b	Zonal	C14	all high mount., large areas >100 000ha	RetezatB.R, RodneiB.R. , BucegiB.R., etc
1 a	***Rhododendro myrtifolii – Piceetum Coldea*** 86 Carpathian mountain pine shrub with spruce and other reduced trees	Timberline b	Zonal	C14/ D55	narrow strip <70 000 ha	Idem
2	***Junipero – Brukenthalietum*** Horv.36 (syn *J. sibiricae, J. nanae* B r.Bl.39 (Mountain dwarf juniper scrub)	Subalp/ Mont b	Zonal	C14	rarely	Idem
3	***Saliceto silesiacae – Alnetum viridis*** Colič et al 62 (Carpathian green alder and montane willow)	Subalp/ Mont b	Alluvia l	C14	rarely	Idem
3 a	***Pulmonario filarskyanae – Alnetum viridis*** Coldea 90 (Carpathian green alder)	Subalp/ Mont b	Alluvia l	C14	very rarely	Rodnei B.R.

No	SHRUBLAND TYPE	Vegetation zone and belts	Vege-tation type	Vegeta-tion map unit.	Widespread	Presence in protected area
4	***Calamagrosti epigeios-Tamaricetum ramosissimi*** Simon Dihoru 63 (syn. *Tamaricetum pallasii* Borza 31) (Pontic tamarisk shrubs)	Plain forest z	Alluvia l	U24a	frequent along the S rivers	Danube Delta B.R., P. N. Balta Brailei
5	***Myricario – Epilobietum*** Aichinger33 (German tamarisk and fireweed)	Hill-plain forest z	Alluvia l	U15,24	sometimes along E,N rivers	Rodnei B.R
6	***Hippophaëtum rhamnoides*** Popescu et al 86 (Sea buckthorn shrub)	Hill-plain forest z	Alluvia l	U 15	sometimes along the rivers	Danube Delta B.R.
7	***Salicetum purpureae*** Wend. Zel 52 (Purple willow shrub)	Hill-plain forest z	Alluvia l	U 19	frequent	-
8	***Salici purpureae – Myricarietum*** Moor 58 (German tamarisk and purple willow)	Hill forest zone	Alluvia l	U 14	sometimes	Rodnei B.R.
9	***Salicetum triandrae*** Malcuit 29 (Almond willow)	Plain forest zone	Alluvia l	U15	sometimes	-
9 a	***Salicetum triandrae*** Malcuit 29 ***amorphosum fruticosae*** Borza 54* (Almond willow)	Plain forest zone	Alluvia l	U15	sometimes	-
10	***Salicetum triandro – viminalis*** (Tx 31) Lohm 52 (Almond willows and common osier shrubs)	Hill-plain forest z	Alluvia l	U 14	rarely, N rivers	Rodnei B.R.

No	SHRUBLAND TYPE	Vegetation zone and belts	Vege-tation type	Vegeta-tion map unit.	Widespread	Presence in protected area
11	***Salicetum cinereae*** Zólyomi 31 (Grey willow shrubs)	Plain forest and steppe	Alluvial	L12a	sometimes	Danube Delta B.R.
12	***Salicetum rosmarinifoliae – Holoschoenetum vulgaris*** Mititelu et al 73 (Dwarf willow on dunes)	Sandy plain	Sand-soil	L12a	rarely	Danube Delta B.R.
13	***Prunetum tenellae*** Soó 46 (Small almond shrub)	Steppe plain	Wet slope	L9	very rarely	-
14	***Caragana mollis, Paliurus spina christi, Jasminus fruticans*** shrublands (termophilous species)	Steppe plain	Wet slope	M4,15	very rarely	Dobrudga areas
	SECONDARY VEGETATION					
15	***Campanulo abietinae-Juniperetum*** Simon 66 (Mountain dwarf juniper with Campanula)	Subalp/Mont	Zonal	D55	rarely	Rodnei B.R.
16	***Crataego - Cerasetum (Prunetum) fruticosae*** Soó (27) 51 (Hawthorn – privet scrub)	Steppe plain	Zonal	G-L-M	rarely	-
17	***Pruno spinosae – Crataegetum*** (Soó27) Hueck 31 (Blackthorn and hawthorn shrubs)	Hill and plain forest	Zonal	G12G8 F 40a L	frequent	-

No	SHRUBLAND TYPE	Vegetation zone and belts	Vege-tation type	Vegeta-tion map unit.	Widespread	Presence in protected area
18	***Syringo - Carpinetum orientalis*** Jakucs 59 (Lilac flower and oriental hornbeam shrub)	Dry hills	Zonal	G11	rarely	-
19	***Coryletum avellanae*** Soó 27 (Hazel nut shrub)	Piedmont, hill	Zonal	F,G	sometimes	-
20	***Eleagnus angustifolia*** shrubland * (Smelling willow)	Costal sand	Alluvial and marine sand	P4	rarely	Black Sea sand

-*shrubs introduced and now sub native
-European vegetation map unit:C14- subalpine shrubs; D55 spruce forest F29 beech forest; F 40..G..oak and common oak and xerophilous oak forest; L2..12, M psamophilous steppe and steppe; U 14 –24 floodplain shrub and forest; -B.R. Biosphere Reservation; P.N. Natural Park

Along the mountain streams grow green alder together with mountain willows in a mixture with tall herbs or other meso-hygrophilous herb species. This association and the next one are arranged on narrow strips along streams, in a mosaic of small areas.

All the other associations shown in the table are developed in the plain and hilly regions and their common feature is a "strip" disposition on small or medium compact areas of distribution but their appearance has different frequency. For example, the most frequent in the southern region of Romanian Plain, but along the rivers is the tamarisk shrub. It colonises alluvial soils and accepts even an alkaline soil with different salt content; on sand it could grow too, as on seashore, but their size is very different.

Myricario-Epilobietum and *Hippophaë rhamnoides* shrublands are spread along the river too but in collinar and piedmont zone, where tamarisk doesn't climb.

All willow shrublands construct some galleries along minor riverbeds; they love wet sites more than the other alluvial shrublands, excepting dwarf willow, which lives on sand protosoils between dunes in dry conditions.

Prunetum tenellae and the other termophilous shrubs have a very small frequency and when they appear in the dessert steppe areas they are represented only by few individuals. Their importance in Romania is only that "they exist", and enrich natural plant diversity of the land.

The secondary vegetation is developed after clearing of forest sometimes just after substitution of pasture. The primary vegetation could replace again the secondary vegetation only without any anthropic influence during a longer or shorter succession. Secondary vegetation will never be in a climax (stabile equilibrium) with the abiotic environment being in a permanent change.

Secondary shrubland at high altitude is noted at timberline – dwarf juniper with campanula- an association, which cover the soil of spruce (*Picea abies*) forest. Their areas are small and not very frequent.

The most frequent areas of secondary vegetation are covered with blackthorn and hawthorn shrubs as well as other *Crataegus* (hawthorn) shrub combinations; they are spread especially in the plain region, in dry zone of steppe and sylvosteppe but in alluvial land appear too. Hazelnut shrubs there are on slope instead of beech and oak forest; *Syringo-Carpinetum* is a rarity, the shrubland is met only in the Danube Gorge, at "Portile de Fier".

The tow shrubs introduced in Romania in forestry culture, *Amorpha fruticosa* and *Eleagnus angustifolia*, were adapted very well at some new sites and are now in expansion. Especially the smelling willow is a very important shrub, because it could use "limit condition" for a wood plant on Black Sea sand – one part of eastern boundary of Romania.

Under natural conditions the shrubs are harvested (cut) irrespective of the species, the inhabitants using them as fire wood and the surrounding land is intensely grazed. Both anthropic activities destroy plants and habitats and our research activities show that even if the shrubs resist as species they are in decline as individuals. Therefore only their inclusion in a Reserve Area could save these habitats.

We can see in Table 1 that most shrubland types are included in some Biosphere Reserves and National Parks. The mountain shrublands are included in Retezat Mountain Biosphere Reserve, Rodnei Biosphere Reserve, Calimani National Park, Bucegi Natural Park and so on. The other shrublands developed in the wetland were recognised in Danube Delta Biosphere Reservation and in Low Danube floodplain, at Balta (Insula) Mica a Brailei – Natural Park too.

The most endangered shrublands are the termophilous shrubland, spread in dry plain in South respectively in Romanian Plain and Dobrudgea Plateau where there are not Parks and where the large agriculture land include these small shrub areas. A special protection is necessary to be initiated for them.

How do all these shrublands look? In the Table 2 we can see some structural characteristics. They are very different: shrubs could be between 0.5 and 7 (10) m tall –the genetics of the dominant species and the environment can influence them too. The height of all shrubs is most frequently between 2.5 and 4 m.

Typical of shrubs is a high density of plants and especially of stems. Many of them are very thorny; *Crataegus monogyna, Hippophaë rhamnoides or Eleagnus angustifolia* are the highest example. Close to them are mountain pine and mountain juniper, coniferous which have evergreen leaves as needles, it being also a protection for a mature – old plant.

Commonly, the bigger shrubs have more levels in their structure development but at the same time the density of steams is more reduced.

Compact shrublands are developed by *Pinus mugo Tamarix ramosissima* and *Hippophaë rhamnoides* . Their height is enough high. However the highest height is met at different willow shrubs. Also, the lowest height is observed at shrubs growing in dry site, in steppe zone on sand soils (*Salix rosmarinifolia, Prunus tenella or Caragana mollis*), excepting *Prunus spinosa* and *Crataegus monogyna,* which could reach at a high size and a high density.

Most of the shrublands are scientifically studied only a little. Their flora and phytosociological association are only partially known. The study of shrubland as ecosystem is not so developed in Romania or in other countries. The lack of economic reasons could be an explanation.

However because of new the concept of "sustainable development", some ecological research has been done. Two shrublands were more studied, in the mountain region – mountain pine and in the plain region - pontic tamarisk and both were a good surprise for the high level of biodiversity and for the complex inside structure of the ecosystem. In most situations, shrubland are ecotone zones and they could be a general determination of high diversity

Table 2. Structure Characteristics of Shrubland Types from Romania

No	SHRUBLAND TYPE	Shrub height (m)	Density	How much are known
1	*Rhododendro myrtifolii – Pinetum mugi* Borza 53 em.Coldea 84 (Carpathian mountain pine shrub)	(0.40) 2 – 3.5	High density 2200shr/ha	Plant biodiversity and ecosystem structure
1 a	***Rhododendro myrtifolii – Piceetum Coldea*** 86 Carpathian mountain pine shrub with spruce and other reduced trees	>2.5	High density	Idem
2	***Junipero –Brukenthalietum*** Horv.36 (syn *J. sibiricae*, *J. nanae* B r.Bl.39 (Mountain dwarf juniper scrub)	0.5 – 1.5	Medium density	Plant biodiversity
3	***Saliceto silesiacae – Alnetum viridis*** Colič et al 62 (Carpathian green alder and montane willow)	4	Medium-low density	Idem
3 a	***Pulmonario filarskyanae –Alnetum viridis*** Coldea 90 (Carpathian green alder)	4	Idem	Idem
4	***Calamagrosti epigeios-Tamaricetum ramosissimi*** Simon Dihoru 63 (syn. *Tamaricetum pallasii* Borza 31) (Pontic tamarisk shrubs)	(0.5) 1.5 - 4 (6.5)	High density 16000- 3800 (200) st/ha	Plant biodiversity and ecosystem structure
5	***Myricario – Epilobietum*** Aichinger33 (German tamarisk and fireweed)	1.5 - 2.5	Medium density	Plant biodiversity
6	***Hippophaëtum rhamnoides*** Popescu et al 86 (Sea buckthorn shrub)	1 - 4 (10)	High density	Idem
7	***Salicetum purpureae*** Wend. Zel 52 (Purple willow shrub)	3 – 5	Medium density	Idem
8	***Salici purpureae – Myricarietum*** Moor 58 (German tamarisk and purple willow)	3 – 5	Medium density	Idem
9	***Salicetum triandrae*** Malcuit 29 (Almond willow)	2 –5 (7)	Medium density	Idem
9 a	***Salicetum triandrae*** Malcuit 29 ***amorphosum fruticosae*** Borza 54* (Almond willow)	2 – 5 (7)	Medium density	Idem
10	***Salicetum triandro – viminalis*** (Tx 31) Lohm 52 (Almond willows and common osier shrubs)	2 - 4 (7)	Medium density	Idem

11	***Salicetum cinereae*** Zólyomi 31 (Grey willow shrubs)	1.5 - 6	Medium density	Idem
12	***Salicetum rosmarinifoliae –Holoschoenetum vulgaris*** Mititelu et al 73 (Dwarf willow on dunes)	1	Low density	Idem
13	***Prunetum tenellae*** Soó 46 (Small almond shrub)	1.5	Low density	Idem
14	***Caragana mollis, Paliurus spina christi, Jasminus fruticans*** shrublands (termophilous species)	1.5 - 3	Few and rare individuals,	Idem
	SECONDARY VEGETATION			
15	***Campanulo abietinae- Juniperetum*** Simon 66 (Mountain dwarf juniper with Campanula)	1 – 1.5 (3)	Medium density	Idem
16	***Crataego - Cerasetum (Prunetum) fruticosae*** Soó (27) 51 (Hawthorn – privet scrub)	0.5 - 1.5	High density	Idem
17	***Pruno spinosae – Crataegetum*** (Soó27) Hueck 31 (Blackthorn and hawthorn shrubs)	(1.5)2.5 – 5.5(10)	High thorny density	Idem
18	***Syringo -Carpinetum orientalis*** Jakucs 59 (Lilac flower and oriental hornbeam shrub)	3 – 7	Medium density	Idem
19	***Coryletum avellanae*** Soó 27 (Hazel nut shrub)	6	Divers density	Idem
20	***Eleagnus angustifolia*** shrubland * (Smelling willow)	3 - 7	High density	no

*shrubs introduced and now subnative

This diversity, exemplified in the case of tamarisk shrublands studied along the river Danube, in Danube Delta and along an inside river Prahova from Romanian Plain, tributary of Danube. There were two distinct situations: - some areas in usual management conditions and areas in protection condition (Table 3, Table 4).

Table 3. Biometrical characteristics of Tamarix ramosissima at investigated areas from Low Danube flood plain

Indici	Insula Mica	Gura Garlutei	Varsat u-ra	Lati-nu	Caraor-man	Suli-na	Sahalin
Shrubs height (m)							
maximum	6.5	2.4	3.44	2.50	2.83	1.5	0.3 (1.5)
mean	4.0	0.96	2.40	1.50	1.53	0.6	0.1(0.8)
Basal diameter (cm)							
maximum	28.9	4.45	5.0	3.1	3.39	3.5	0.6(4.5)
mean	9.4	2.56	3.5	1.7	1.81	1.6	0.2(2.0)
Vegetation cover %							
Shrub layer	95	60	60	80	50	90	10(60)
Herbaceous layer	52-88	90	80	70	90	80	5(20)
Shrubland age (year)	27	27	15	10	15	15	1
Investigated area (sq m)	12000	2000	10000	5000	10000	2000	100

Table 4 Tamarix ramosissima shoot basal diameter (cm)and height in the two sites of Low Prahova floodplain

Site	Shrub diameter cm	Shoot height (m)		
		Mean	Highest	Lowest
Pisculeşti	5.0	1.36	3.53	0.36
Dridu	3.9	0.96	2.29	0.36

Seven sites are studied on the Danube; only the first is included in National Park Balta (Insula) Mica a Brailei, 25 years old and partially protected (far from settlements, in natural flooded conditions, grazed only by semiwild farmer's animals. On the Prahova river biometrical data show a smaller size of shrublands at Pisculesti and at Dridu, but just so they are over 2 m height.

The best development of shrubs is observed at Insula Mica a Brailei where the plants are "near trees" as height and as basal diameter of stems. At the same age the shrubland from Gura Garlutei situated near settlements is more reduced as size and as cover of land.

A special case is met at Sahalin, a small island in Black Sea, near Danube Delta, where shrubland grew higher and well developed on the marine sand but an occasional storm destroyed it completely.

The biodiversity of plants reaches 197 species on Danube and more than 230 on Prahova tamariks shrublands, with an active change from a site to the other and an active dynamics from a year to the other (annual number in a site is not more than 60 species). In Insula Mica a Brailei, the canopy of shrubs is full of bird nests (3 / 10sq m); the ground covers hole of different animals as fox (Vulpes vulpes) or snake and hare (Lepus europeus) shelter. The invertebrate fauna is very rich, as species as well as individuals in canopy layer, at herb level or in soil. Some graphics could be seen in Pauca Comanescu et al., 1997, 2000.

CONCLUSION OR PERSPECTIVE

The structure of shrubland – with its distribution in dense strips alternating with grassland, with a medium height between 2 and 4 m, and with a high density of stems, could be, we suppose, a good camouflage system. Something more, the location of shrublands is on special lands, at high altitude, along the rivers, along the seashore; it is isolated and easier to control.

The necessity to have an undisturbed place (without stem cutting or animal grazing) is common for shrubland development for a normal wild life protected by nature conservation actions as well as for certain army activities such as the protection of some objectives . These areas could have a double use and a double protection. Of course, these lands couldn't be used as army training area or for a polluting activity, but in consent it could be used for a camouflage aim.

REFERENCES

Environmental Romanian Strategy- Romanian Position Document, Chapter 22, Environmental Protection, 2000

Engagement Declaration of National Defence Ministry, 2001

Coldea Gh., *Muntii Rodnei. Studiu geobotanic*, Bucuresti, Ed. Acad. 1990

Doniță N., Ivan Doina, Coldea Gh., Sanda V., Popescu A., Chifu Th.,

Paucă – Comănescu Mihaela, Mititelu D., Boşcaiu N., *Vegetația României* , Bucureşti, Ed. Tehnica Agricolă, , 1992

Paucă-Comănescu M., Vasiliu-Oromulu L., Şerbănescu Gh., Tăcină A., Falcă M.,Honciuc V., Purice D., Sterghiu Cl., Cociu M., Tatole V., Ceianu I., Dumitru L., Faghi M., Gomoiu I., Vicol C., Ecosystemic characterization of *Tamarix ramosissima* shrublands from Small

Island of Braila (Danube flood plain). Proceedings of the Institute of Biology, Annual Scientific Session, 1, Bucharest,147 - 179, 1997

Paucă-Comănescu M., Şerbănescu Gh., Spatial and temporal dynamics of the species presence of *Tamarix ramosissima* shrublands growing in the Danube floodplain and Delta. Rev. Roum.,- Biol. Veget.,43, 2, 121 - 142, 1998

Paucă – Comănescu M., Onete M, Tăcină A. Species diversity of primary producers of *Tamarix* shrublands, in Prahova and Teleajen low floodplain, Proceedings of the Institute of Biology, III, 99 – 117, Ann. Ses. 2000, (2001)

Paucă – Comănescu M., Tăcină A., Vasiliu – Oromulu L., Honciuc V., Falcă M., Onete M, Purice D., Stănescu M, Blujdea V., Ionescu M., Structure of the main biocenotic components in *Tamarix* shrublands of low Prahova and Teleajen floodplain, Rev. Roum. Biol., s. Biol Veget., 45, 2, 155-179, 2000, (2002)

GPSR Compliance

The European Union's (EU) General Product Safety Regulation (GPSR) is a set of rules that requires consumer products to be safe and our obligations to ensure this.

If you have any concerns about our products, you can contact us on ProductSafety@springernature.com

In case Publisher is established outside the EU, the EU authorized representative is:

Springer Nature Customer Service Center GmbH
Europaplatz 3
69115 Heidelberg, Germany

Batch number: 10824476

Printed by Printforce, the Netherlands